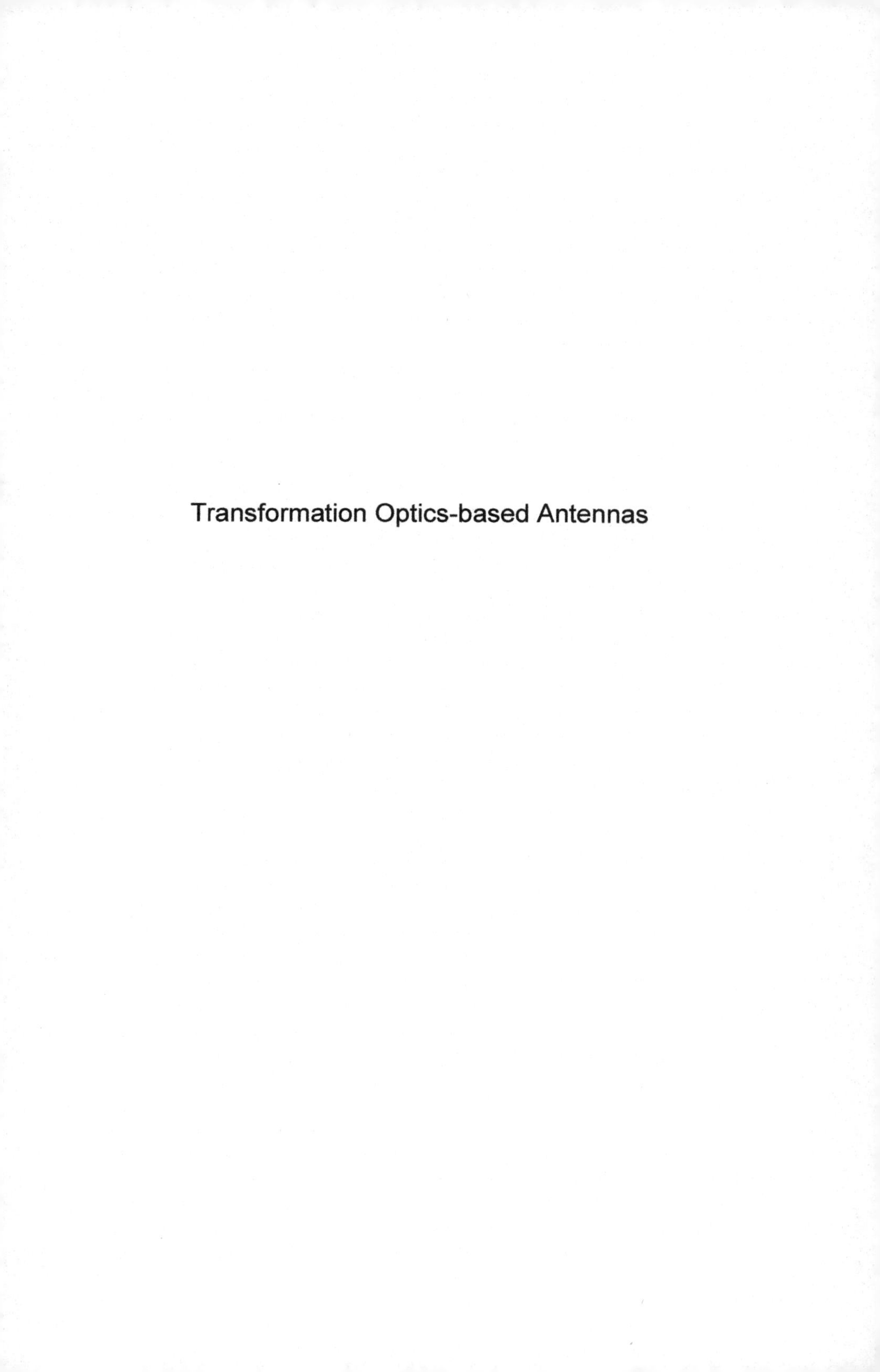

Transformation Optics-based Antennas

Metamaterials Applied to Waves Set

coordinated by
Frédérique de Fornel and Sébastien Guenneau

Transformation Optics-based Antennas

Shah Nawaz Burokur
André de Lustrac
Jianjia Yi
Paul-Henri Tichit

First published 2016 in Great Britain and the United States by ISTE Press Ltd and Elsevier Ltd

ISTE Press Ltd
27-37 St George's Road
London SW19 4EU
UK

www.iste.co.uk

Elsevier Ltd
The Boulevard, Langford Lane
Kidlington, Oxford, OX5 1GB
UK

www.elsevier.com

British Library Cataloguing-in-Publication Data
A CIP record for this book is available from the British Library
Library of Congress Cataloging in Publication Data
A catalog record for this book is available from the Library of Congress
ISBN 978-1-78548-197-0

Printed and bound in the UK and US

Contents

Preface

The subject of this book is an interesting research topic called transformation optics (or transformation electromagnetics) and its application to the control of the path of electromagnetic waves through an association with another interesting concept known as metamaterial engineering technology. Although the fundamental principle of deformation of wave path in an inhomogeneous medium has been known for decades, it was only in 2006 that the concept of transformation optics was established to materialize space deformation to give light such a desired path. Such a concept is, then, able to allow the design of novel and unimaginable electromagnetic and optical devices for various functionalities.

This book focuses on the theoretical tools defining transformation optics concept. We address the origin of the concept by analyzing Fermat's principle. We, then, discuss the two main methods of transformation that allow the design of devices. We detail the basic approaches and the methods to design practical applications of transformation optics concepts for beginners in the field such as engineers, Masters and PhD students. Through antenna applications, we aim to provide the readers with the whole process of designing a device based on transformation optics, right from theoretical formulations to implementation and subsequent experimental validation.

The purpose of this book is two-fold: to demonstrate that transformation optics is not only a powerful theoretical way to design unbelievable and novel devices such as invisibility cloaks, but also a realization tool of microwave devices with unusual properties that are difficult to achieve with conventional methods. In particular, we detail the design of anisotropic materials used in these applications. We show that the main criticism regarding resonant metamaterials, i.e. their small bandwidth, can be overcome. We show also that it is possible to vary several electromagnetic parameters simultaneously using the metamaterial technology. We also show that 3D manufacturing can be used efficiently to realize low-cost fast prototyping of electromagnetic devices for electromagnetic radiation control. Secondly, we imagine that this book can be a source of inspiration and a practical tool for engineers and researchers to develop new types of unusual electromagnetic devices.

The book is organized into two chapters after the introduction. The introduction presents the basic principles of the transformation optics concept. Two types of transformation are presented: coordinate transformation and space transformation. Implementations using metamaterials are also discussed. Chapter 1 focuses on coordinate transformation to design devices capable of modifying the electromagnetic appearance of a radiating source. Transformation of a directive radiation pattern into an isotropic one and vice-versa through a space stretching and compression, respectively, and the possibility to create multiple beams are studied. Chapter 2 deals with devices designed using the space transformation concept. Quasi-conformal transformation optics (QCTO) is applied for the design of lenses either to compensate for the phase shift created by the conformation of an array of sources or to steer a beam to an off-normal direction. Materials are engineered through 3D printing and prototypes presenting a variation in electromagnetic parameters are fabricated and tested to validate the proposed lenses.

In summary, this book presents theoretical concepts as well as practical methods to ensure effective implement transformation optics based devices. Such realizable designs open the way to new types of electromagnetic devices for applications in various domains such as telecommunications, aeronautics and transport.

Shah Nawaz BUROKUR,
André DE LUSTRAC, Jianjia YI,
Paul-Henri TICHIT
March 2016

Introduction

I.1. Where does transformation optics (TO) come from?

Pierre de Fermat, in a letter written in 1662, gave the principle that leads to geometrical optics [TAN 91]. The path of light between two points is stationary. Fermat's principle states that light follows the Extremum Optical Path, which is often the shortest one, between two points in space. Mathematically, the optical path s is defined infinitesimally as the product of the refractive index n and an elementary distance dl:

$$s = \int ndl \qquad [I.1]$$

We note that in the case of a constant index, for example, that of a homogeneous and isotropic medium, the optical path is proportional to the path of the light beam. In the well-known case of a flat Euclidean geometry, the shortest path is the straight line, which is not the case in a curved space. When the index is no longer constant, for example, in a non-homogeneous medium, the shortest path is not a straight line, but a curve. This is the case of the mirage effect in summers where the optical index of the air layers above a hot road varies with temperature. In this case, we observe a curvature of light that gives the impression that the road is covered with water.

Figure I.1 illustrates the deformation of a light ray when the space in which the ray propagates is distorted (Figure I.1(a) and (b)). Hence, it can be seen that a variation of optical index can be equivalent to a distortion of space, and it can be concluded that a material supporting such optical index variation can simulate a distorted space. This is due to the relation between the metrics of space and the solution of Maxwell's equations in this space. J.B. Pendry and U. Leonhardt, in their papers published in 2006, referred to the invariance of Maxwell's equations in a distorted space [PEN 06, LEO 06a]. J.B. Pendry concluded that it is possible to hide a region of space by a conformation of light rays around this region, as shown in Figure I.1(c).

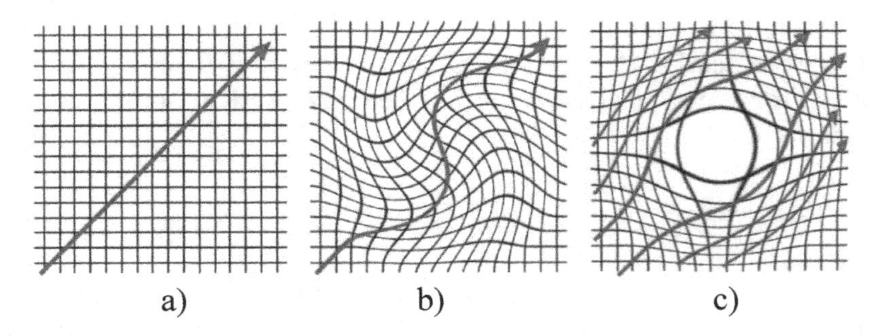

Figure I.1. *a) propagation of a light ray in a non-distorted space; b) propagation of the same light ray in a distorted space; c) cloaking of a region of space for incoming light rays*

To explain the conformation of light rays, let us consider a function of the form:

$$M = \int F\left[x, y, y'(x)\right] dx \qquad [I.2]$$

We can show by Euler-Lagrange equations that an extremum of this function is solution of:

$$\frac{\partial F[x,y,y'(x)]}{\partial y}-\frac{d}{dx}\left(\frac{\partial F[x,y,y'(x)]}{\partial y'}\right)=0 \qquad [I.3]$$

We can then understand why light travels in a straight line in a medium where the index is constant. Indeed, the function to minimize is:

$$L_{AB}=\int_A^B ndl \overset{2D}{=} n\int_A^B \sqrt{dx^2+dy^2}=n\int_A^B \sqrt{1+y'^2(x)}dx \qquad [I.4]$$

with,

$$F[x,y,y'(x)]=\sqrt{1+y'^2(x)} \qquad [I.5]$$

So, from equation [I.3], we obtain $y''(x)=0$ or more precisely $y(x)=ax+b$, which is the equation of a straight line. We can, therefore, note that it is possible to know the path of light from an index profile and vice versa. More generally, from Maxwell's equations and the choice of an electromagnetic field, the eikonal equation can be established in the regime of geometrical optics (small wavelength compared with the characteristic dimensions of the environment). This equation, which can also be deduced from the optical path is written as $\left(\vec{\nabla}L\right)^2=n^2$ and allows obtaining the fundamental equation of light rays in an isotropic inhomogeneous medium by derivation:

$$\frac{d}{ds}\left(n\frac{d\vec{r}}{ds}\right)=\vec{\nabla}n \qquad [I.6]$$

In the case of anisotropic media, the calculation of the trajectory of the light rays is done from the dispersion equation directly established for a field polarization and the Hamilton equations. Now that we know the physics of light rays, the question that might arise is: what happens to Fermat's principle if we bend space? In the next section, we show that Fermat's principle is preserved, although this implies constraints on transformations. The conservation of Fermat's principle by invariance of space change is referred to as conformal transformation. We will see also that such transformation generates material with achievable parameters as the anisotropy is weak.

I.2. Conformal transformations

In this section, we show the interest of the conservation of Fermat's principle by coordinate transformation. Imagine that we illuminate our empty space by several light sources. Then we transform our space, which is a Cartesian grid, by coordinate transformation such that the light rays follow a different path. To simplify our problem, let us be in a two-dimensional (2D) space and set our transformation as follows:

$$\begin{cases} x' = f(x,y) \\ y' = g(x,y) \end{cases} \quad [I.7]$$

Is Fermat's principle always obeyed by this general transformation?

From equation [I.7], we are able to rewrite the principle in the new coordinate system:

$$L'_{AB} = \int_{A'}^{B'} n'(x',y')dl' = \int_{A'}^{B'} n'(x',y')\sqrt{dx'^2 + dy'^2} \quad [I.8]$$

And by derivation of relation [I.7], we have

$$dx'^2 = \left(\frac{\partial f(x,y)}{\partial x} dx + \frac{\partial f(x,y)}{\partial y} dy \right)^2$$
$$dy'^2 = \left(\frac{\partial g(x,y)}{\partial x} dx + \frac{\partial g(x,y)}{\partial y} dy \right)^2 \qquad [I.9]$$

Which implies that,

$$dx'^2 + dy'^2 = \left(\frac{\partial f(x,y)}{\partial x} dx + \frac{\partial f(x,y)}{\partial y} dy \right)^2 + \left(\frac{\partial g(x,y)}{\partial x} dx + \frac{\partial g(x,y)}{\partial y} dy \right)^2$$

$$dx'^2 + dy'^2 = \left(\left(\frac{\partial f}{\partial x} \right)^2 + \left(\frac{\partial g}{\partial x} \right)^2 \right) dx^2 + \left(\left(\frac{\partial f}{\partial y} \right)^2 + \left(\frac{\partial g}{\partial y} \right)^2 \right) dy^2 + 2 \left(\left(\frac{\partial f}{\partial x} \right)\left(\frac{\partial f}{\partial y} \right) + \left(\frac{\partial g}{\partial x} \right)\left(\frac{\partial g}{\partial y} \right) \right) dxdy$$

By identification, we finally obtain:

$$\frac{\partial f(x,y)}{\partial x} = \frac{\partial g(x,y)}{\partial y}$$
$$\frac{\partial f(x,y)}{\partial y} = -\frac{\partial g(x,y)}{\partial x} \qquad [I.10]$$

with the additional condition of transformation of the index:

$$n'^2(x',y') \left(\left(\frac{\partial f}{\partial x} \right)^2 + \left(\frac{\partial f}{\partial y} \right)^2 \right) = n^2(x,y) \qquad [I.11]$$

When equations [I.10] and [I.11] are fulfilled, the transformation [I.7] is called conformal transformation. Two possible interpretations can be given: first, the classical interpretation consists of being in the primed coordinate system and thereby to assign the curved path of the light rays to the $n'(x',y')$ index. The second interpretation is to consider two empty spaces and to assign the curvature of the light rays to a different space metric, i.e. of a curved space. Conformal transformation is a typical case since the condition [I.10] is not often obtained. The interesting aspect of this transformation is the conservation of Fermat's principle and the possibility to have an isotropic inhomogeneous index as given by equation [I.11]. The index profile to achieve is very simple since the problem is not addressed by the permittivity and permeability that give very often complex and impractical anisotropies.

We extend this conformal transformation to the wave nature of light and consider the Helmholtz's propagation equation. It is written as:

$$\left(\Delta + k^2\right)\phi = 0 \text{ with } k^2 = \frac{\omega^2}{c^2}n^2 \qquad [I.12]$$

Considering the coordinate transformation of [I.7], the Laplacian operator is transformed into:

$$\Delta = \frac{\partial^2}{\partial x^2} + \frac{\partial^2}{\partial y^2} = \left(\frac{\partial}{\partial x} + i\frac{\partial}{\partial y}\right)\left(\frac{\partial}{\partial x} - i\frac{\partial}{\partial y}\right) \qquad [I.13]$$

However, the two partial derivatives can be written as:

$$\begin{aligned} \frac{\partial}{\partial x} &= \frac{\partial}{\partial x'}\frac{\partial f(x,y)}{\partial x} + \frac{\partial}{\partial y'}\frac{\partial g(x,y)}{\partial x} \\ \frac{\partial}{\partial y} &= \frac{\partial}{\partial x'}\frac{\partial f(x,y)}{\partial y} + \frac{\partial}{\partial y'}\frac{\partial g(x,y)}{\partial y} \end{aligned} \qquad [I.14]$$

Using equation [I.10], the first term of equation [I.13] is written as

$$\left(\frac{\partial}{\partial x}+i\frac{\partial}{\partial y}\right)=\left(\frac{\partial g(x,y)}{\partial y}-i\frac{\partial f(x,y)}{\partial y}\right)\left(\frac{\partial}{\partial x'}-i\frac{\partial}{\partial y'}\right), \quad [I.15]$$

and the second term as,

$$\left(\frac{\partial}{\partial x}-i\frac{\partial}{\partial y}\right)=\left(\frac{\partial g(x,y)}{\partial y}+i\frac{\partial f(x,y)}{\partial y}\right)\left(\frac{\partial}{\partial x'}+i\frac{\partial}{\partial y'}\right) \quad [I.16]$$

Thus, the Laplacian operator is:

$$\begin{aligned}\Delta &= \left(\frac{\partial}{\partial x}+i\frac{\partial}{\partial y}\right)\left(\frac{\partial}{\partial x}-i\frac{\partial}{\partial y}\right)\\ &= \left(\left(\frac{\partial g(x,y)}{\partial y}\right)^2+\left(\frac{\partial f(x,y)}{\partial y}\right)^2\right)\Delta'\\ &= \left(\left(\frac{\partial f(x,y)}{\partial x}\right)^2+\left(\frac{\partial f(x,y)}{\partial y}\right)^2\right)\Delta'\end{aligned} \quad [I.17]$$

with $\Delta'=\frac{\partial^2}{\partial x'^2}+\frac{\partial^2}{\partial y'^2}$.

The Helmholtz's equation becomes:

$$\begin{aligned}&\left(\Delta+\frac{\omega^2}{c^2}n^2\right)\phi=0\\ &\left(\left(\frac{\partial f(x,y)}{\partial x}\right)^2+\left(\frac{\partial f(x,y)}{\partial y}\right)^2\right)\Delta'\phi+\frac{\omega^2}{c^2}n^2\phi=0\\ &\Delta'\phi+\frac{\omega^2}{c^2}\frac{n^2}{\left(\left(\frac{\partial f(x,y)}{\partial x}\right)^2+\left(\frac{\partial f(x,y)}{\partial y}\right)^2\right)}\phi=0\end{aligned} \quad [I.18]$$

According to equation [I.11], we have:

$$\left(\Delta' + \frac{\omega^2}{c^2} n'^2\right)\phi = 0 \qquad [I.19]$$

The Helmholtz's equation is, therefore, invariant under conformal transformation and the waves solutions in the initial space are transformed in the same way as light rays. However, working with conformal transformations requires more constraints in the targeted transformation and often generates a very complex index profile. So, it is easier to work without constraints with other types of transformations. This requires being in curved spaces with a particular geometry.

Now that the concept of transforming light rays has been introduced, we are able to introduce the concept of transformation optics in Chapter 1. Two types of transformations will be detailed: coordinate transformation and space transformation. The structures used presently to calculate the material parameters for the transformations will also be presented.

Chapter 2 will focus on antenna applications derived from coordinate transformation. The concept of transformation of electromagnetic sources including miniaturization of radiating sources will be presented. Transformation media of directive and isotropic antennas will thus be calculated. For practical applications, we will first stress on metamaterial engineering for specific permittivity and permeability values.

Chapter 3 will deal with lens applications calculated from space transformations. The designed lenses allow for controlling the path of electromagnetic waves for restoring in-phase emissions and steering radiated beam of an antenna. Since space transformation allows calculating the transformed media from isotropic materials, implementation from all-dielectric materials will be presented.

1

Transformation Optics Concept: Definition and Tools

1.1. State of the art on metamaterials

The exciting field of metamaterials has attracted the interest of researchers from all across the world over the past few years. With roots in the foundations of optics and electromagnetics, researchers are exploring the underlying physics of metamaterials. Electromagnetic metamaterials are artificially structured materials composed of periodic arrays of sub-wavelength resonating structures, scatterers or apertures, whose electric or magnetic response provides the ability to engineer dielectric or magnetic properties that do not exist in naturally occurring materials. The dielectric or magnetic properties can be tailored by changing the geometrical parameters of the constituent structure of the metamaterial.

Metamaterials has covered a wide variety of topics with a rich history of scientific exploration. A metamaterial is a microstructured material that is patterned to achieve a desired interaction with some wave-type phenomena, and is usually either acoustic or electromagnetic. The electromagnetic metamaterials studied and applied in this book mainly concern the engineering of unusual permittivity and permeability values such as negative or less than unity.

The recent surge in interest in metamaterials, however, has been fueled by experimental and theoretical demonstrations of physical phenomena, resulting from this patterning, which mimic materials that are not available in nature. Figure 1.1 shows the most well-known metamaterial implementation, the emblematical electromagnetic cloaking device [SCH 06a].

Figure 1.1. *Two versions of an electromagnetic cloaking device using metamaterials designed by the D.R. Smith group at Duke University*

The principle of metamaterials relies on sub-wavelength structures engineering. Through this engineering, the metamaterial has the functionality of an electric or magnetic particle that behaves in the same way as an atom or molecule of continuous media. Metamaterials composed of particle structures, such as split ring resonators (SRRs) [PEN 99], electric-LC (ELC) resonators [SCH 06b] and cut wires [KAF 05, SHA 07], the effective ε and μ of the equivalent medium can achieve both positive and negative values across certain frequency bands. Although the concept of negative permittivity has been accepted and understood for a long time, negative permeability is a less familiar phenomenon typically not found in natural materials generally, except for ferrite materials.

In naturally occurring materials with a positive index of refraction, solutions of the Maxwell equations yield propagating electromagnetic waves in which the phase and energy velocities are in the same direction. To explain why this is true, imagine a uniform plane wave propagating in an unbounded medium so that no boundary conditions are applied to the wave. Taking the direction of propagation of this plane wave with angular frequency ω to be in the $+z$ direction and an electric field polarized along the x direction, we can express the electric and magnetic fields as:

$$\begin{aligned} E(z,t) &= E_0 e^{j(\omega t - kz)}\hat{x} \\ H(z,t) &= \frac{E_0}{\eta} e^{j(\omega t - kz)}\hat{y} \end{aligned} \qquad [1.1]$$

where the wave number $k = k_r + jk_i$ is complex, with k_i contributing to attenuation along the direction of propagation. The power flow in such a medium is given by the time averaged Poynting vector.

$$S = \frac{1}{2}\Re\left(E \times H^*\right) = \frac{1}{2}\Re\left(\frac{k_z E_x^2}{\omega\mu}\right)\hat{z} \qquad [1.2]$$

As presented in Figure 1.2, in most natural materials, ε and μ are positive, and this forces k_z to assume a continuum of real values. Thus, S maintains a non-zero value, and energy is free to propagate in the +z direction.

The case when both μ and ε are simultaneously negative is not classically considered. In such case, the material has a refractive index n, which is negative. The wave number

$k_z = \frac{\omega}{c}\sqrt{\mu_r \varepsilon_r}$ becomes negative, while the power flow S flows in the $+z$ direction. Thus, the power flow becomes anti-parallel with the phase velocity, and because of this property, negative index materials are referred to as "left-handed" materials since a simultaneously negative ε and μ makes the vector triplet [E, H, k] left-handed. "Left-handed" materials are called so because the wave vector is antiparallel to the Poynting vector, seemingly violating the right-hand rule.

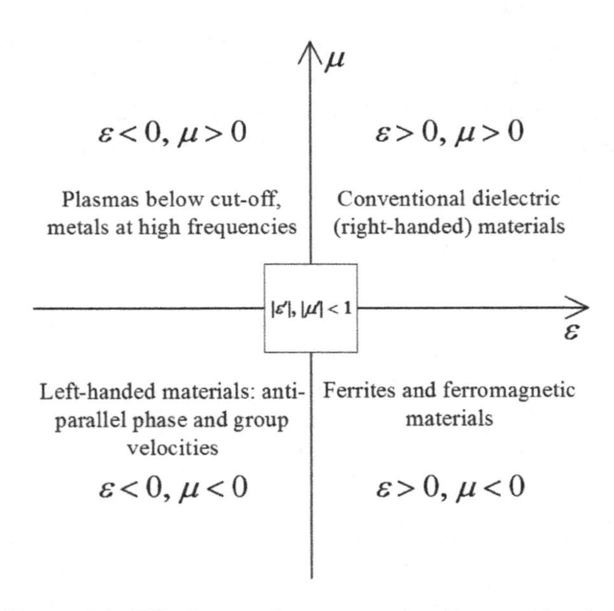

Figure 1.2. *Effective medium parameter diagram showing the four quadrants of material parameters ε and μ*

Figure 1.3 shows the comparison between an incident ray propagating into a positive index and a negative index medium. In a positive index medium, waves with a larger wave vector than in free space are evanescent: rather than oscillating in free space, they decay exponentially. In a negative index medium, however, the amplitudes of these evanescent waves grow exponentially, allowing the transmission of information about features on a much finer scale than is usually available using conventional optical elements.

A further significant advantage of the manual patterning of a material is that it can be made inhomogeneous and anisotropic, i.e. the properties of the material can change as a function of position and the field polarization. This allows for the curving of light ray trajectories throughout a material and can lead to a significantly increased phase space in which optical devices can be designed.

Lots of applications based on metamaterials have been proposed since J.B. Pendry introduced SRRs and wire structures [PEN 98] to realize negative permeability and permittivity medium respectively. For instance, they were used to create the first negative index material [SHE 01].

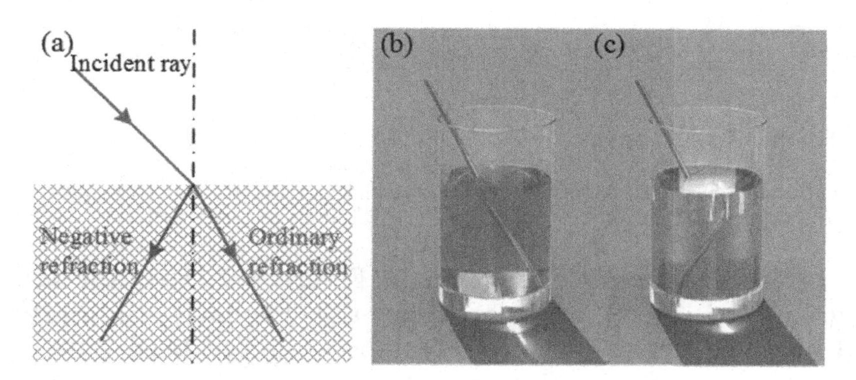

Figure 1.3. *Comparison of refraction of an incident ray propagating into a positive and negative index material [DOL 06]*

Numerous applications of metamaterials have been proposed, particularly for the antenna domain. However, left-handed or negative index metamaterials are not generally used. Rather, controllable zero or positive index are of great interests for antennas. For example, in

Figure 1.4, a zero index metamaterial is introduced for directive emission by using the ultra-refraction effect [ENO 02]. According to Snell-Descartes laws, when a light ray goes through an interface from a zero index medium to a positive index medium, the transmitted ray is normal to the interface in the positive index medium. This is called the ultra-refraction effect. In this application, under proper conditions, the metallic metamaterial grid embedding the source is engineered to produce a zero index. Thus, the energy radiated by the embedded source is concentrated in a narrow cone in the surrounding medium, as illustrated in Figure 1.4(a). The refracted rays in vacuum are almost in the same direction around the normal and thus a directive emission is obtained.

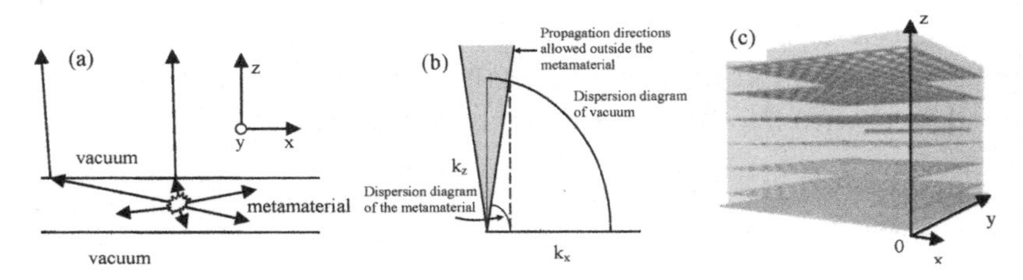

Figure 1.4. *a) Geometrical interpretation of the emission of a source inside a slab of a low index metamaterial (not negative); b) construction in the reciprocal space; c) schematic representation of the structure [ENO 02]*

A second example is given in Figure 1.5. The research group led by D. Werner from the Pennsylvania State University together with researchers at Lockheed Martin Commercial Space Systems have succeeded in enabling specific device performance over usable bandwidths by employing dispersion engineering of metamaterial properties [LIE 11]. In particular, they have designed metamaterial liners that considerably improve the performance of conventional horn antennas over greater than an octave bandwidth with negligible losses. The artificial metamaterial is used to taper the aperture field to the E-plane so as to reduce the sidelobes level (SLL). In this application, the metamaterial liner inside the horn transforms the metallic surface into a "soft" surface presenting a high impedance Z_{TM} for the transverse magnetic mode and a low impedance Z_{TE} for transverse electric mode. Figure 1.5(b) shows the frequency evolutions of both impedances, with Z_0 being the vacuum impedance.

In Figure 1.6, a new type of compact flexible anisotropic metamaterial coating was proposed to greatly enhance the impedance bandwidth of a quarter-wave monopole to over an octave [JIA 11a]. The metamaterial coating has a high effective permittivity for the tensor component oriented along the direction of the monopole. By choosing the appropriate radius and tensor parameter of the metamaterial coating, another resonance at a higher frequency can be efficiently excited without affecting the fundamental mode of the monopole. Additionally, the similar current distributions on the monopole at both resonances make stable radiation patterns possible over the entire frequency band.

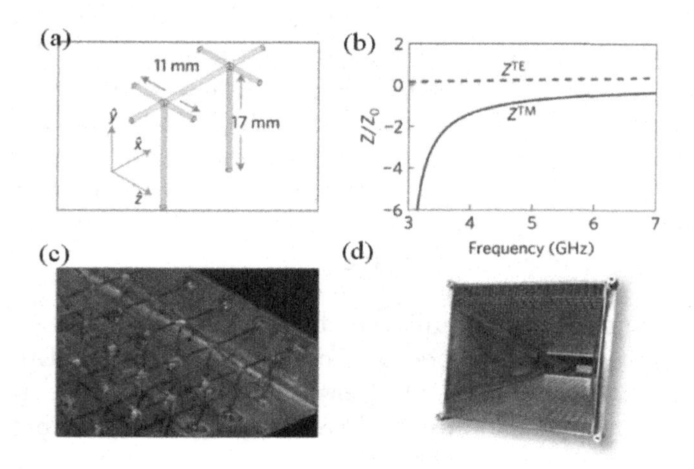

Figure 1.5. *a) Metamaterial liner geometry; b) normalized TE- and TM-polarized surface impedances; c) photograph of the fabricated metamaterial; d) photograph of the metamaterial-implemented horn antenna [LIE 11]*

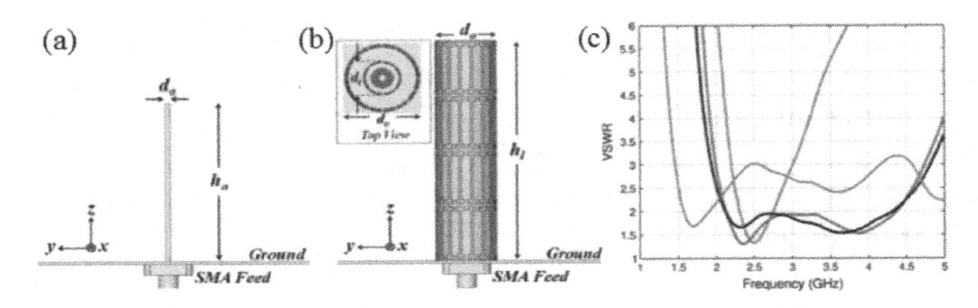

Figure 1.6. *a) Monopole antenna alone; b) monopole antenna with ultrathin flexile anisotropic metamaterial coating; c) simulated and measured impedance matching [JIA 11a]. The curves show firstly the narrow response of the monopole alone and secondly the calculated and measured broad responses of the monopole coated with the metamaterial*

Other antennas based on the use of gradient metamaterial-based lenses have been recently presented in literature. In the first work reported in the use of gradient index (GRIN) metamaterials, the authors made use of SRRs with non-uniform dimensions and showed a deviation of an electromagnetic wave to validate the concept [SMI 05]. For example, a gradient-index lens was utilized to steer the radiated beam of a horn antenna [MA 09a]. Also, gradient index in negative index metamaterials was designed [GRE 05, DRI 06]. In [DRI 06], T. Driscoll and colleagues presented a radial gradient-index lens with an index of refraction ranging from −2.67 (edge) to −0.97 (center), as shown in Figure 1.7. The authors showed that the lens can produce field intensities at the focus that are greater than that of the incident plane wave.

To avoid using volumetric bulky metamaterials, GRIN structures were developed in guided configuration [LIU 09c]. Indeed, guide mode is possible through the use of complementary resonators. These types of resonators have been introduced in [FAL 04] for the design of planar homogeneous metasurfaces by applying Babinet’s principle to conventional metallic metamaterial resonators. A planar homogeneous near-zero permittivity metasurface was experimentally validated in waveguided configuration for electromagnetic tunneling [LIU 08]. These realizations, though interesting, were limited by their narrow bandwidth, due to the intrinsic resonant nature of the metamaterial structure used. GRIN metamaterials using non-resonant cells for wideband operations were then developed [LIU 09a]. The main idea behind such non-resonant elements was to make use of $\lambda/4$ or $\lambda/2$ resonant elements. Such resonators cannot be considered homogeneous at the resonance frequency due to their relatively large dimensions. However, at frequencies much lower than that of the resonance an array of such resonators behave as an effective medium with electromagnetic parameters presenting low losses and dispersion.

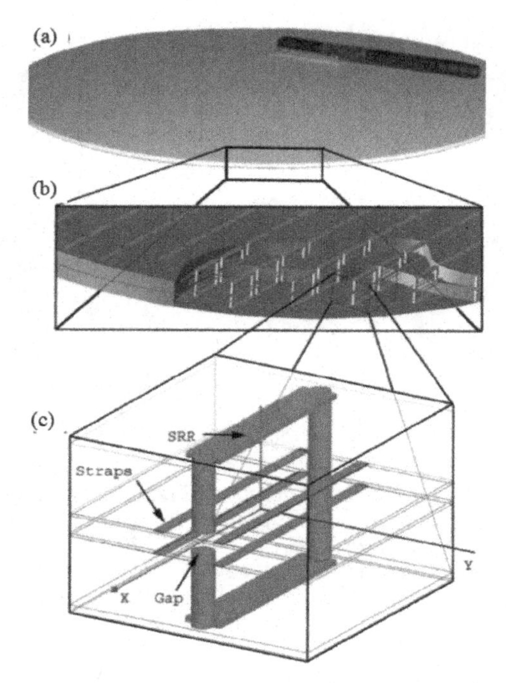

Figure 1.7. *Diagrams showing a) picture of a lens disk, b) details of the constituting metamaterial cells, c) unit cell of the metamaterial used [DRI 06]*

Broadband GRIN lenses have then been developed for directive antennas [MA 09b, MA 10a, MA 10b, MA 11, CHE 09a, PFE 10, KUN 10, DHO 12a, DHO 13a, DHO 13b]. Figure 1.8 shows, for example, the design of a planar 2D Luneburg lens antenna operating in a waveguided configuration. Each cell is made of a complementary closed resonator (CCR), which has an effective index determined by its geometrical dimensions. Varying these dimensions, we can design a 2D GRIN lens as in the Figure 1.8. In this realization, the lens is divided into 7 zones with a radially effective index varying from 1 at the periphery to 1.4 at the center. The lens is illuminated by a Vivaldi-type antenna producing a wideband directive emission.

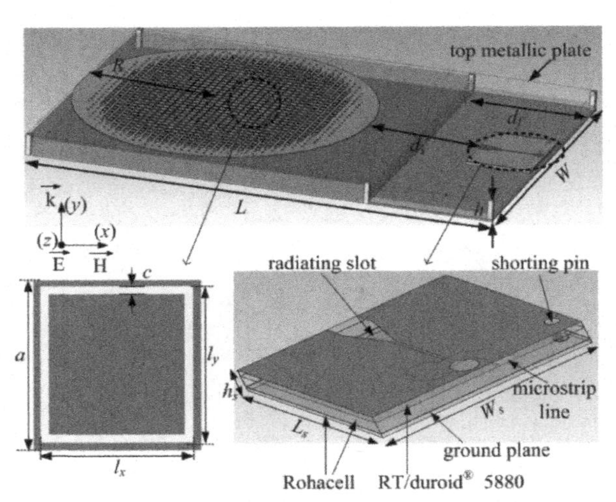

Figure 1.8. *Luneburg lens antenna composed of a planar metamaterial-based lens and a Vivaldi antenna feed placed in a quasi-TEM parallel-plate waveguide [DHO 12a]. In the CCR cell: a = 3.333 mm, c = 0.3 mm and l_x and l_y are variable parameters. In the Vivaldi feed: L_S = 26.5 mm, W_S = 35 mm and h_S = 5.2 mm. In the whole antenna system: L = 190 mm, W = 130 mm, h = 11 mm, d_f = 50 mm, d_S = 51.5 mm and R = 56 mm*

The simulated and measured radiation patterns are presented in Figure 1.9. A highly directive emission is observed at all tested frequencies.

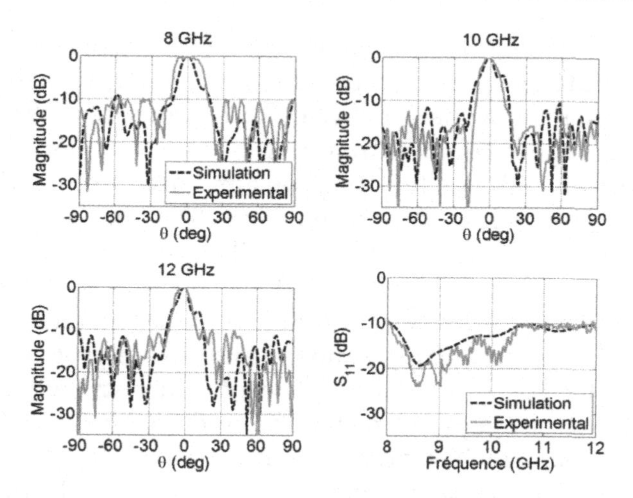

Figure 1.9. *Simulated and measured far-field radiation patterns and return losses of the Luneburg lens-antenna [DHO 12a]*

These examples show that we are able to engineer, in a relatively easy way, the electromagnetic properties of metamaterials so as to obtain new artificial materials with a negative, null or positive index, by adjusting the geometrical dimensions of metallic inclusions on dielectric substrates judiciously.

Numerous other works can be found in the literature concerning metamaterial-based and metamaterial-inspired antennas. The antenna characteristics targeted from the use of metamaterials are typically size reduction, multiple resonances, impedance matching and high directivity.

1.2. Transformation optics

Transformation optics (TO) is a powerful and conceptual technique for the design of devices which leverage complex materials. Based on the invariance of Maxwell's equations during a change of coordinate system, it offers a great potential for the design of devices with unbelievable properties. It enables altering the field distribution by applying material parameters which cause light to "behave" as if it was in a transformed set of coordinates [WAR 96]. Moreover, the combination of TO and structured metamaterials allow for the design and realization of material properties generated by TO concept. From metamaterials, it is possible to tailor the desired spatial distribution of permittivity and permeability, therefore, offering the possibility to guide and control the path of light by such space engineering. A new vision to electromagnetism is brought by TO. Analogue to general relativity, where time and space are considered to be one and can be modified by energy and mass, TO shows that is possible to apply a deformation in space to specify a desired path to light. Hence, general relativity finds an application in novel optical systems based on TO where light is guided by engineered materials in a desired and controlled manner.

As stated in the introduction, light generally travels along the shortest optical path. When the index is a complex tonsorial distribution, light does not travel in a straight path. We can think that this is only true within the geometrical optics limit, i.e. for small wavelengths as compared to geometrical dimensions. In fact, the spatial coordinate transformation gives the appropriate permittivity and permeability tensors due to the invariance of Maxwell's equations, which makes it possible to control light at any scale, from macroscopic to microscopic. By creating the permittivity and permeability distributions, space can be deformed at will giving the opportunity to modify light path in a fantastic way at will.

Based on this technique, novel complex electromagnetic systems have been explored over the last ten years, including:

– invisibility cloaks [SCH 06a, CAI 07, GRE 08, YAN 08, LI 08, GAI 08, LIU 09b, GAB 09, VAL 09, KAN 09, ERG 10, FAR 08];

– electromagnetic concentrators [JIA 08a, LUO 08a, RAH 08a];

– electromagnetic field rotators [CHE 07, CHE 09b];

– wormholes [GRE 07, ZHA 09];

– black holes (absorbers) [NAR 09, CHE 10];

– transformation plasmonics [KAD 14];

– waveguide transitions and bends [RAH 08b, RAH 08c, HUA 09, ROB 09];

– illusion devices [LAI 09, LI 10, JIA 11b, JIA 13];

– focusing lenses and antennas [GUE 05, KWO 08, LEO 08, KUN 08, LUO 08b, JIA 08b, RAH 08b, ALL 09, KON 07, KUN 10, MA 09b, MA 10b, POP 09, TUR 10, QUE 13].

Although the concept was already known for many years [TAM 24, PLE 60], the introduction of TO in 2006 helped to bring a correspondence between geometry and materials. In this way, the material can be considered as a new geometry and information about space coordinate transformation is given by the properties of the material. Based on the invariance of Maxwell's equations, the arbitrary control of the field electromagnetic is made possible by introducing a specific space coordinate transformation that transforms an initial space into another one possessing the desired virtual properties. Among the classes of transformation in the literature, several possibilities are available for the design of electromagnetic structures. For example, continuous transformations were introduced by J.B. Pendry [PEN 06] for the first invisibility cloak, which led to anisotropic and heterogeneous permittivity and permeability tensors. Continuous transformations offer a substantial advantage in their generality of applications. Such a transformation has been used in many cases as mentioned above. In parallel,

U. Leonhardt [LEO 06a] proposed the concept of conformal transformation where the transformation obeys Fermat's principle and enables the design of devices by the use of isotropic dielectric media [LAN 09, HAN 10, SCH 10, TUR 10]. The main drawback of this type of transformation is that it often requires mathematical requirements that are too complex for design systems. Following this idea, quasi-conformal transformation has emerged [LI 08, LIU 09b, GAB 09, VAL 09, ERG 10, CHA 10], where a slight deformation in the transformation minimizes the anisotropy of the material and allows for an approximation of the device with an isotropic medium. Other theoretical works have appeared from space–time transformation [LEO 00, BER 08, CRU 09, THO 11], creating a link with cosmology and celestial mechanics [GEN 09]. Finally, source transformation allowing for the design of devices with a source included in the transformed space has also been proposed [LUO 08b, LUO 08c, ALL 09, POP 09].

Practical applications of TO go well beyond the invisibility cloak and this is particularly the subject of this book. In fact, the theory allows controlling light in an ultimate way by providing permittivity and permeability distributions. Using these distributions, the optical space is shaped to obtain the desired path of light. The challenge is, then, to find the correct transformation so that realization with metamaterials is made possible. This science of light by transformation optics associated with the use of metamaterials paves the way for new and fascinating applications that could not be conceived otherwise.

To provide the reader with the background information necessary to understand the rest of the book, this chapter is devoted to a complete introduction of different kinds of transformations. Transformation optics relies on what is termed the "form-invariance" of Maxwell's equations, which means that transformation optics provide a method to modify the field by changing the variables in the Maxwell's equations. This chapter begins with a derivation of this property with the familiar form of Maxwell's equations. It is then followed by a description of the consistent manner in which transformation optical calculations may be performed for arbitrary geometries. Transformations based on Laplace's equations are also introduced. Finally, some applications will serve as examples to help understanding the described principles.

1.2.1. *Coordinate transformation*

In this section, the transformation from Cartesian coordinates to other orthogonal coordinate systems is introduced by the statement of the media equations. Several applications of such coordinate transformations will also be introduced later. We start from Maxwell's equations to explain the principles and then focus on writing out the material parameters in a very explicit way for future studies by means of numerical finite element methods. Coordinate transformation is achieved by redefining the constituent material parameters. Maxwell's equations are written such that the differential form of the equations is invariant under coordinate transforms

$$
\begin{aligned}
&\nabla \times H = \varepsilon \frac{\partial E}{\partial t} + J \\
&\nabla \times E = -\mu \frac{\partial H}{\partial t} \\
&\nabla \cdot \varepsilon E = \rho \\
&\nabla \cdot \mu H = 0
\end{aligned}
\qquad [1.3]
$$

These equations are rewritten in covariant notation in Cartesian coordinates. The invariance form is a property of the differential equations. The covariant notation is used for compactness.

$$
\begin{aligned}
&\varepsilon^{ijk}\partial_j H_k = \varepsilon^{ij}\frac{\partial E_j}{\partial t} + J^i \\
&\varepsilon^{ijk}\partial_j E_k = -\mu^{ij}\frac{\partial H_j}{\partial t} \\
&\partial_i \varepsilon^{ij} E_j = \rho \\
&\partial_i \mu^{ij} H_j = 0
\end{aligned}
\qquad [1.4]
$$

where, i, j and k are the indices that take values from 1 to 3. We note that a particular coordinate is identified by x^i, for example (x, y, z). And ∂_i is shortly written as $\partial/\partial x^i$. The completely anti-symmetric Levi-Cevita tensor is expressed by ε^{ijk}.

We consider the properties of these equations to lie under a coordinate transformation between x and $x^{i'}$, which is the transformed coordinate (x', y', z'). E and H are transformed as:

$$
\begin{aligned}
&E_{j'} = A^{i}_{j'} E_i \\
&E_i = A^{i'}_{i} E_{i'}
\end{aligned}
\qquad [1.5]
$$

where $A^{i'}_{i}$ is the Jacobian matrix and is given by

$$
A^{i'}_{i} = \frac{\partial x^{i'}}{\partial x^{i}} \qquad [1.6]
$$

After several steps of derivations and simplifications, the transformed equations can be written as

$$
\begin{aligned}
&\varepsilon^{i'j'k'}\partial_{j'} H_{k'} = \frac{A^{i'}_{i} A^{j'}_{j}}{\left|A^{i'}_{i}\right|}\varepsilon^{ij}\frac{\partial E_{j'}}{\partial t} + \frac{A^{i'}_{i}}{\left|A^{i'}_{i}\right|} J^i \\
&\varepsilon^{i'j'k'}\partial_{j'} E_{k'} = -\frac{A^{i'}_{i} A^{j'}_{j}}{\left|A^{i'}_{i}\right|}\mu^{ij}\frac{\partial H_{j'}}{\partial t} \\
&\partial_{i'}\frac{A^{i'}_{i} A^{j'}_{j}}{\left|A^{i'}_{i}\right|}\varepsilon^{ij} E_{j'} = \frac{\rho}{\left|A^{i'}_{i}\right|} \\
&\partial_{i'}\frac{A^{i'}_{i} A^{j'}_{j}}{\left|A^{i'}_{i}\right|}\mu^{ij} H_{j'} = 0
\end{aligned}
\qquad [1.7]
$$

where $\left|A^{i'}_{i}\right|$ is the determinant of the Jacobian matrix.

When we compare Maxwell's equations before and after a coordinate transformation, we can find that there exist some transformation rules of the material parameters, current and charge that will ensure the transformed equations are in an identical form as to before transformation. With a proper choice of materials and sources, one can arrive at an identical set of equations to those which are found under an arbitrary coordinate transformation. In other words, if we choose a certain material whose properties are as below:

$$\begin{aligned} \varepsilon^{i'j'} &= \frac{A_i^{i'} A_j^{j'}}{\left|A_i^{i'}\right|} \varepsilon^{ij} \\ \mu^{i'j'} &= \frac{A_i^{i'} A_j^{j'}}{\left|A_i^{i'}\right|} \mu^{ij} \\ J^{i'} &= \frac{A_i^{i'}}{\left|A_i^{i'}\right|} J \\ \rho' &= \frac{\rho}{\left|A_i^{i'}\right|} \end{aligned} \quad [1.8]$$

then, the transformed Maxwell's equations can be rewritten as

$$\begin{aligned} \varepsilon^{i'j'k'} \partial_{j'} H_{k'} &= \varepsilon^{i'j'} \frac{\partial E_{j'}}{\partial t} + J^{i'} \\ \varepsilon^{i'j'k'} \partial_{j'} E_{k'} &= -\mu^{i'j'} \frac{\partial H_{j'}}{\partial t} \\ \partial_{i'} \varepsilon^{i'j'} E_{j'} &= \rho' \\ \partial_{i'} \mu^{i'j'} H_{j'} &= 0 \end{aligned} \quad [1.9]$$

Now, we continue with the transformation between two Cartesian coordinate systems:

$$\varepsilon_C^{\ i'j'} = \frac{A_i^{i'} A_j^{j'} \varepsilon_C^{\ ij}}{\left|A_i^{i'}\right|}$$
$$\mu_C^{\ i'j'} = \frac{A_i^{i'} A_j^{j'} \mu_C^{\ ij}}{\left|A_i^{i'}\right|} \qquad [1.10]$$

where ε_C^{ij} and μ_C^{ij} are the components of the permittivity and permeability tensors in the original Cartesian coordinate. $\varepsilon_C^{i'j'}$ and $\mu_C^{i'j'}$ are the components of the permittivity and permeability tensors of the transformed media in the new Cartesian coordinate. The subscript *C* stands for Cartesian coordinate system.

For the transformation between any two kinds of orthogonal coordinate systems, such as between cylindrical coordinate system and spherical coordinate system, we define the original coordinate system as (u, v, w) and the transformed coordinate system as (u', v', w'), to distinguish with the typical Cartesian coordinate expression (x, y, z). The mapping $(u,v,w) \Leftrightarrow (u',v',w')$ can be written as

$$\begin{cases} u' = u'(u,v,w) \\ v' = v'(u,v,w) \\ w' = w'(u,v,w) \end{cases} \qquad [1.11]$$

The transformation media equations in the orthogonal coordinate (defined by the subscript *O* in the following equations) can be expressed as,

$$\varepsilon_O^{\ k'l'} = T_k^{k'} T_l^{l'} \varepsilon_O^{\ kl} / \det(T_k^{k'})$$
$$\mu_O^{\ k'l'} = T_k^{k'} T_l^{l'} \mu_O^{\ kl} / \det(T_k^{k'}) \qquad [1.12]$$

with $T_k^{k'} = (h_{u^{k'}} \partial u^{k'}) / (h_{u^k} \partial u^k)$, h_{u^k} is defined in the line element expression in the orthogonal coordinate in the original space.

In a Cartesian coordinate system, the line element in the original space is $ds^2 = dx^2 + dy^2 + dz^2$. While in the orthogonal coordinate in the original space, the line element is

$$\begin{aligned} ds^2 &= h_u(u,v,w)^2 du^2 + h_v(u,v,w)^2 dv^2 + h_w(u,v,w)^2 dw^2 \\ &= h_u{}^2 du^2 + h_v{}^2 dv^2 + h_w{}^2 dw^2 \end{aligned} \quad [1.13]$$

For different coordinate systems, of course, h_u, h_v and h_w have different expressions, which are given for Cartesian coordinates (x, y, z), cylindrical coordinates (ρ, ϕ, z) and spherical coordinates (r, θ, ϕ) as shown in Table 1.1.

Cartesian coordinates $ds^2 = dx^2 + dy^2 + dz^2$	$\begin{cases} h_u = 1 \\ h_v = 1 \\ h_w = 1 \end{cases}$
Cylindrical coordinates $ds^2 = d\rho^2 + \rho^2 d\phi^2 + dz^2$	$\begin{cases} h_u = 1 \\ h_v = \rho \\ h_w = 1 \end{cases}$
Spherical coordinates $ds^2 = dr^2 + r^2 d\theta^2 + r^2 \sin^2\theta d\varphi^2$	$\begin{cases} h_u = 1 \\ h_v = r \\ h_w = r\sin\theta \end{cases}$

Table 1.1. *Expressions for h_u, h_v, and h_w in different coordinate systems*

So, the Jacobian transformation matrix in the orthogonal coordinate system is:

$$T = \begin{bmatrix} \frac{h_{u'}}{h_u}\frac{\partial u'}{\partial u} & \frac{h_{u'}}{h_v}\frac{\partial u'}{\partial v} & \frac{h_{u'}}{h_w}\frac{\partial u'}{\partial w} \\ \frac{h_{v'}}{h_u}\frac{\partial v'}{\partial u} & \frac{h_{v'}}{h_v}\frac{\partial v'}{\partial v} & \frac{h_{v'}}{h_w}\frac{\partial v'}{\partial w} \\ \frac{h_{w'}}{h_u}\frac{\partial w'}{\partial u} & \frac{h_{w'}}{h_v}\frac{\partial w'}{\partial v} & \frac{h_{w'}}{h_w}\frac{\partial w'}{\partial w} \end{bmatrix} \quad [1.14]$$

For example, if the transformation takes place between two cylindrical coordinate systems $(\rho,\phi,z)\Leftrightarrow(\rho',\phi',z')$, the Jacobian transformation matrix takes the form:

$$T=\begin{bmatrix} \frac{\partial\rho'}{\partial\rho} & \frac{1}{\rho}\frac{\partial\rho'}{\partial\phi} & \frac{\partial\rho'}{\partial z} \\ \rho'\frac{\partial\phi'}{\partial\rho} & \frac{\rho'}{\rho}\frac{\partial\phi'}{\partial\phi} & \rho'\frac{\partial\phi'}{\partial z} \\ \frac{\partial z'}{\partial\rho} & \frac{1}{\rho}\frac{\partial z'}{\partial\phi} & \frac{\partial z'}{\partial z} \end{bmatrix} \quad [1.15]$$

Normally, the parameters obtained above will be assigned as properties to the transformed media and simulated by numerical finite element methods. For instance, the COMSOL Multiphysics finite element-based electromagnetics solver can be used. So, the required parameters should be expressed in the form of (x', y', z'):

$$\begin{aligned} \varepsilon_C^{\ i'j'} &= R\varepsilon_O^{\ k'l'}R^T / \det(R) \\ \mu_C^{\ i'j'} &= R\mu_O^{\ k'l'}R^T / \det(R) \end{aligned} \quad [1.16]$$

where, the rotation matrix R is given as

$$R=\begin{bmatrix} \frac{1}{h_{u'}}\frac{\partial x'}{\partial u'} & \frac{1}{h_{v'}}\frac{\partial x'}{\partial v'} & \frac{1}{h_{w'}}\frac{\partial x'}{\partial w'} \\ \frac{1}{h_{u'}}\frac{\partial y'}{\partial u'} & \frac{1}{h_{v'}}\frac{\partial y'}{\partial v'} & \frac{1}{h_{w'}}\frac{\partial y'}{\partial w'} \\ \frac{1}{h_{u'}}\frac{\partial z'}{\partial u'} & \frac{1}{h_{v'}}\frac{\partial z'}{\partial v'} & \frac{1}{h_{w'}}\frac{\partial z'}{\partial w'} \end{bmatrix} \quad [1.17]$$

For cylindrical and spherical coordinate systems, the expressions of matrix R are presented in Table 1.2.

Cylindrical coordinates	$R=\begin{bmatrix} \cos\phi' & -\sin\phi' & 0 \\ \sin\phi' & \cos\phi' & 0 \\ 0 & 0 & 1 \end{bmatrix}$
Spherical coordinates	$R=\begin{bmatrix} \sin\theta'\cos\varphi' & -\cos\theta'\cos\varphi' & -\sin\varphi' \\ \sin\theta'\sin\varphi' & -\cos\theta'\sin\varphi' & \cos\varphi' \\ \cos\theta' & \sin\theta' & 0 \end{bmatrix}$

Table 1.2. *Expressions of rotation matrix R in different coordinate systems*

Now, let's take an example to see how this concept works in a device design. In 2008, D.R. Smith proposed to use coordinate transformation to design a device which can bend an electromagnetic beam by a certain angle [RAH 08c]. The schematic principle is presented in Figure 1.10. The variable k in the formulation represents a scale factor and α is the bending angle.

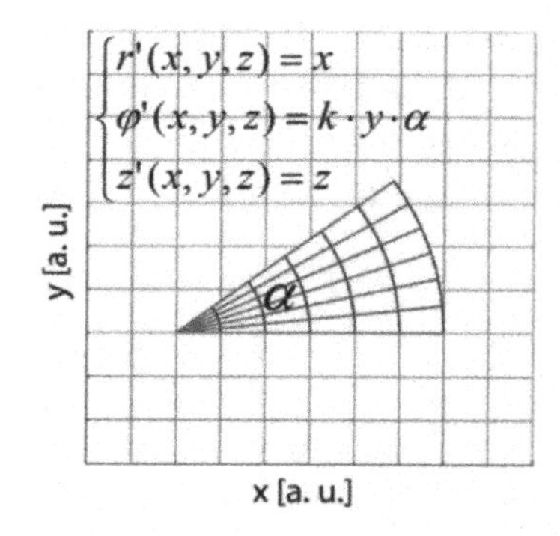

Figure 1.10. *Illustration of the embedded coordinate transformation for bending waves ($\alpha = \pi/5$) proposed in [RAH 08c]*

As can be seen from the formulation, the transformation is between an original Cartesian coordinate system and a transformed cylindrical coordinate system. To give a more detailed explanation, the x-axis is transformed to the radial distance r', and the y-axis is transformed to a scaled azimuth axis. Following the recipe discussed above, one obtains the permittivity and permeability tensors of the beam bend expressed in cylindrical coordinate system.

$$\varepsilon^{i'j'} = \mu^{i'j'} = \begin{bmatrix} \gamma / r & & \\ & r / \gamma & \\ & & \gamma / r \end{bmatrix} \qquad [1.18]$$

with $\gamma = 1/k\alpha$. Note that r' was substituted by r for possible simulations in COMSOL Multiphysics.

The polarization of the electric field was chosen to lie in the z-direction normal to the xy-plane of the wave propagation. The frequency considered in the simulations was 8.5 GHz. However, it should be noted that the transformation-optical approach is valid for any arbitrary frequency. The dark grey lines represent the direction of the power flow. Figure 1.11 shows the spatial distribution of the electric field for a $\pi/2$-bend. In Figure 1.11(a), the material properties were calculated for γ = 0.02 whereas in Figure 1.11(b) the tensors were obtained for γ = 0.3. As can be seen, each material realization bends the incoming light by $\alpha = \pi/2$.

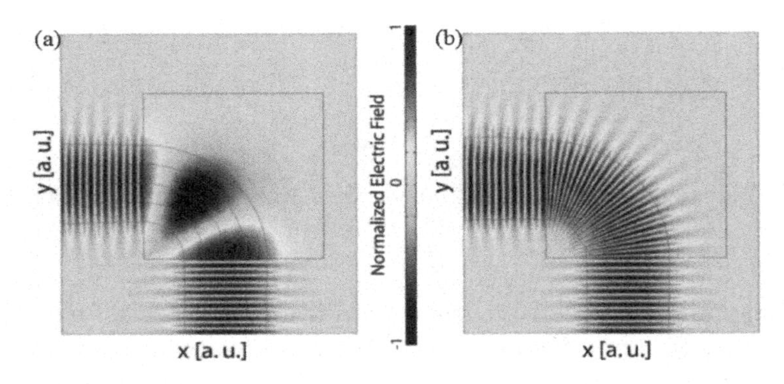

Figure 1.11. *Distribution of the electric field component normal to the plane of propagation for $\alpha = \pi/2$ with a) γ = 0.02 and b) γ = 0.3 [RAH 08c]*

1.2.2. *Space transformation*

The specific coordinate transformation is used to determine the distribution of constitutive parameters discussed in section 1.2.1. These determined parameters are assigned to a TO device so as to obtain functional physical field distributions. However, the field distributions within the volume of the device is of no consequence in most of the cases: only the fields on the boundaries of the device are relevant, since the function of most optical devices is to match a set of output fields on one port or aperture to a set of input fields on another port or aperture. From the TO point of view, device functionality can be determined by the properties of the coordinate transformation at the boundaries of the domain. Since there is an infinite number of transformations that have identical behavior on the boundary, there is considerable freedom to find a transformation which is close to optimal in the sense that it maximizes a desired quantity, such as isotropy.

1.2.2.1. *Transformations based on Laplace's equation with Dirichlet boundary conditions*

As mentioned in equation [1.10], the permittivity ε' and permeability μ' in the transformed space can be expressed as $\varepsilon' = \mu' = \eta' = \frac{A\eta A^T}{|A|}$, whre A is the Jacobian transformation tensor, which characterizes the geometrical variations between the original space Ω and the transformed space Ω'. The determination of the matrix A is the crucial point for designing transformation media. A can be obtained from the point to point mapping defined by coordinated transformation formulations. A can also be obtained from the mapping between two domains by specifying proper boundary conditions.

Here, we take an arbitrary shaped cloak as example to introduce the transformation based on Laplace's equation with only Dirichlet boundary conditions [LEO 06a, LEO 06b, HU 09].

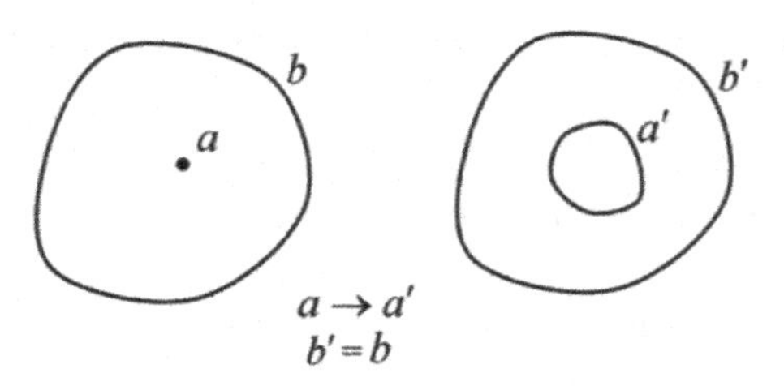

Figure 1.12. *The scheme of constructing an arbitrary cloak [HU 09]*

Figure 1.12 shows the schematic principle to construct an arbitrary shaped cloak. Suppose we have an original region enclosed by the outer boundary represented by b, and inside of this region, we define a point denoted by a. An arbitrary cloak can be constructed by enlarging the point a to an inner boundary a', while keeping the outer boundary of the region fixed ($b = b'$). This condition can be expressed by $U'(a) = a'$ and $U'(b) = b'$, where the operator U' is the new coordinate for a given point during the transformation. Now, the problem is how to determine the deformation field $\partial U'/\partial x$ within the cloak layer enclosed by the inner and outer boundaries and for a specific transformation. The commonly used operator U' for designing a cloak is a linear transformation [PEN 06]. For example, the radial displacement of a spherical cloak is assumed to be a linear relation $r' = (b' - a')r/(b' + a')$. However, for an arbitrary cloak, it is very difficult to express analytically the boundaries a' and b', and the calculation of deformation field is usually very complicated.

To ensure a cloak without reflection, the internal deformation field of the cloak layer must be continuous and wave impedance must be constant along the structure. The deformation tensor is calculated by the partial derivative of displacement with respect to the original coordinate. Therefore, the displacement fields must be smooth enough. It is known that the Laplace's equations with Dirichlet boundary conditions will always give rise to harmonic solutions [COU 89]. This suggests that the displacement field inside of the cloak layer can be calculated by solving Laplace's equations $\Delta U' = 0$ with the boundary conditions $U'(a) = a'$ and $U'(b) = b'$. To eliminate the singular solution of Laplace's equations, we can use the inverse form of the Laplace's equations as

$$\left(\frac{\partial^2}{\partial x_1'^2}+\frac{\partial^2}{\partial x_2'^2}+\frac{\partial^2}{\partial x_3'^2}\right)U_i = 0, \qquad i = 1, 2, 3 \qquad [1.19]$$

where, U_i denotes the coordinates in the original space. It should be noted that in this inverse form, the corresponding Dirichlet boundary conditions are $U(a') = a$ and $U(b') = b$.

After solving the equation [1.19] by a PDE solver which can be found commercially, we are able to obtain $\partial x_j/\partial x'_i$, which means that we can obtain the matrix A^{-1}. Then, A can be calculated and so does the permittivity and permeability.

The two-dimensional arbitrary cloak is verified numerically as presented in Figure 1.13. This concept of transformation provides complex parameters generally which are often too high to achieve. But from the field distribution, the cloaking device performs a good functionality.

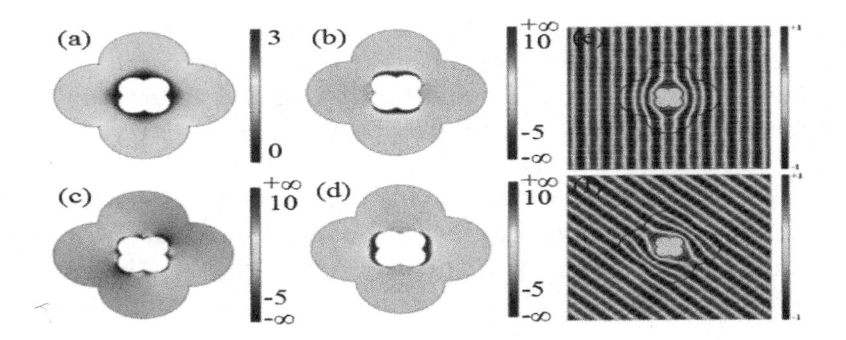

Figure 1.13. *2D arbitrary cloak proposed in [HU 09]. Material parameters a) ε_{zz}, b) μ_{xx}, c) μ_{xy} and d) μ_{yy}. Electric field distribution for incident wave e) horizontally f) 45°*

1.2.2.2. *Conformal mapping*

An invertible mapping is referred to as conformal mapping, if it preserves angles. We start discussing the conformal mapping by a review of the 2D mapping between (*x*, *y*) and (*x*', *y*') systems of the form $X' = [x'(x,y), y'(x,y)]$. The permittivity tensor can be written as:

$$\varepsilon' = \frac{\varepsilon}{|A|}\begin{bmatrix} x'^2_x + x'^2_y & x'_x y'_x + x'_y y'_y & 0 \\ x'_x y'_x + x'_y y'_y & y'^2_x + y'^2_y & 0 \\ 0 & 0 & 1 \end{bmatrix} \quad [1.20]$$

Ideally, the constitutive tensors only have diagonal components, this implies that $x'_x y'_x = -x'_y y'_y$. Since $\varepsilon_{xx} \neq \varepsilon_{yy}$, we can find that

$$\begin{aligned} x'_x &= y'_y \\ x'_y &= -y'_x \end{aligned} \quad [1.21]$$

These equations are the well-known Cauchy-Riemann equations that define conformal maps. We can now find that the coordinates X' satisfy the vector form of Laplace's equation:

$$\nabla^2 x' = 0 \qquad [1.22]$$

This leads to the simple material parameters that are given as:

$$\varepsilon = \mu = \begin{bmatrix} 1 & 0 & 0 \\ 0 & 1 & 0 \\ 0 & 0 & |A|^{-1} \end{bmatrix} \qquad [1.23]$$

TO media that are of the form of equation [1.23] are often described as "isotropic" and "all-dielectric". The full transformation can be obtained from an isotropic dielectric material, if the electric field is polarized along the z-axis (TM_z polarization). Finding conformal mappings analytically while solving the Cauchy-Riemann differential equations explicitly is a challenge. For example, the analysis of mapping between $z = x + iy$ and $z' = x' + iy'$ is very complex. However, the limitations of conformal maps arise when we truncate our transformation domain. In other words, once boundary conditions are applied to the governing differential equations, it is not always possible to achieve conformal modules M, the aspect ratio of the differential rectangle corresponding to a set of orthogonal coordinates, equal to unity, and hence the Cauchy-Riemann equations cannot be applied universally to all geometries.

1.2.2.3. *Quasi-conformal transformation optics (QCTO)*

As discussed previously, much of the functionality of TO is determined by the transformation at the boundary of the domain. For instance, it might be required that our transformation does not introduce reflections or change the direction of a wave entering or exiting our transformed domain [RAH 08b]. These conditions introduce additional restrictions to the transformation. The most straightforward way to satisfy these conditions is to stipulate that the coordinates are the same as free space on the boundary of the transformed domain, for instance the Dirichlet boundary conditions.

An extra constraint concerns the fact that each side of the physical domain should also be mapped to the same corresponding boundary in the virtual domain and this limits the scope of conformally equivalent domains severely. For example, let's think about the mapping between two quadrilateral domains. Once the sides of the quadrilateral domain have been specified, the region can only be mapped to another quadrilateral that shares the same conformal module M. Another concern relates to the boundary conditions directly. While Dirichlet boundaries are ideal for most purposes, they may be incompatible with the requirement of orthogonality at all points in the mapped domain. If we specify $x'(x)$ and M at the boundary simultaneously, the problem becomes over-determined and we are not guaranteed that the mapping will be orthogonal at the boundary [THO 99]. Therefore, a combination of Dirichlet and Neumann boundaries is required to fix the geometry of the transformed domain and maintain orthogonality on the boundary [LAN 14].

We now discuss about the mapping between two domains that do not share the same conformal module. As shown in Figure 1.14, the physical domain is a rectangle domain with a bump on the bottom boundary whose conformal module is M. We intend to map this physical domain to the rectangle virtual domain whose conformal module is 1 (left-hand side of Figure 1.14).

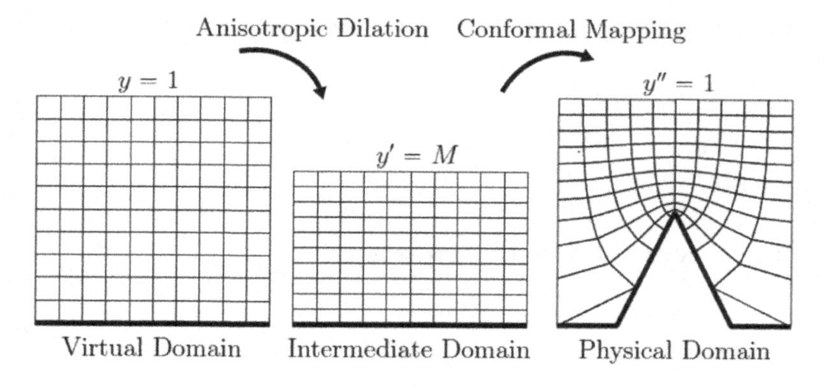

Figure 1.14. *Depiction of the quasi-conformal mapping using intermediate transformations [LAN 14]. Lines of constant x and y, (virtual domain coordinates), are shown in each domain. The thick black line represents a PEC boundary in each domain*

To compensate for the mismatch in conformal modules, we map the virtual domain first to an intermediate domain having the same conformal module as the physical domain. The Jacobian matrix A_c of the conformal transformation between the intermediate and the physical domains is assumed to be obtained. The simplest way to do this is with a uniform dilation of the form $y' = My$. This intermediate domain can then be mapped onto the physical domain with a conformal transformation, such that the mapping between the final transformed and initial coordinates is written as $X'' = \left[x''(x'(x,y), y'(x,y)), y''(x'(x,y), y'(x,y))\right]$

The combined dilation and conformal mapping produce material tensors of the form:

$$\varepsilon'' = \mu'' = Diag\left[M^{-1}, \quad M, \quad (M|A_c|)^{-1}\right] \qquad [1.24]$$

The conformal module provides an immediate measure of the anisotropy of the TO medium, as well as its required magnetic response, since $M=\sqrt{\mu_x/\mu_y}$ [LAN 14]. We have:

$$\begin{aligned} Mx^{"}_x &= y^{"}_y \\ My^{"}_x &= -x^{"}_y \end{aligned} \qquad [1.25]$$

The inverse form of this equation is

$$\begin{aligned} x_{x"} &= My_{y"} \\ My_{x"} &= -x_{y"} \end{aligned} \qquad [1.26]$$

We find that we recover the Laplacian vector for the inverse problem

$$\nabla^2_{x'} x = 0 \qquad [1.27]$$

The following steps are the same as we have discussed previously in the part of the transformation based on Laplace's equation with Dirichlet boundary conditions.

We see that the cost of quasi conformal (QC) mapping is immediately clear: the in-plane material tensors elements are no longer equal to each other. However, it is generally the case that small deformations of space create small perturbations to the conformal module of the physical domain. Li and Pendry [LI 08] suggested that the "small" anisotropy can be ignored in this case. The geometric average of these quantities is simply

$$n=\sqrt{\varepsilon_z}=(M|A_c|)^{-1/2} \qquad [1.28]$$

Since the virtual and physical domains are no longer of the same size, the resulting distributions of material parameters are only approximately correct, and will introduce a number of aberrations. Despite its limitations, the QCTO method has found many applications. For example, the QCTO method has been applied to flatten conventional dielectric lens-antennas and parabolic reflector-antennas without a significant loss in performance [TAN 10, KON 07, MEI 11]. The quasi-isotropic properties can then further be implemented by all-dielectric materials which benefit from a wide frequency band of operation.

1.3. Metamaterial engineering

While QCTO allows for implementing devices with all-dielectric materials having electric responses, coordinate transformation often requires materials with either electric and/or magnetic response. This section is, therefore, dedicated to material engineering, with explanations on how to obtain permittivity and permeability values.

In most cases, coordinate transformation optics based devices are implemented through the use of metamaterial resonators. Such resonators must be small enough compared to the wavelength so that a bulk medium composed of such resonators can be considered homogeneous. The dimension of a typical metamaterial unit cell is around $\lambda/10$ where λ is the free space wavelength. Having a homogeneous medium, then, implies that we are able to extract the effective material parameters, which are the permittivity ε and the permeability μ, of the medium.

The material parameters are calculated by using an inversion method as described in [WEI 74, NIC 70, SMI 02]. From the scattering matrix S, we are able to identify its impedance Z and index n. The electric permittivity is then given as $\varepsilon = n/Z$, and the magnetic permeability is $\mu = nZ$. Both n and Z, and therefore ε and μ, are complex functions that depend on frequency and must satisfy certain requirements based on causality. For example, in the case of passive materials, Re(Z) and Im(n) must be positive.

1.3.1. *Electric resonators*

The most common electric resonators are cut wires [KUN 10], I-shaped resonators [LIU 09b, DHO 12b] and ELC resonators [SCH 06b] (see Figure 1.5). While I-shaped and ELC resonators have subwavelength dimensions, it is not generally the case for cut wires. So, when dealing with cut wires, one should be careful about the correct frequency-band where the structure can be considered to be homogeneous in order to retrieve correct material parameters.

1.3.2. *Magnetic resonators*

The most well-known artificially magnetic resonator is the split ring resonator (SRR) introduced by J.B. Pendry in 1999, and illustrated in Figure 1.16(a) [PEN 99]. Such a resonator produces a magnetic resonance when the magnetic field H is polarized along the axis of the rings. Therefore, this resonator operates under a parallel-to-plane propagation. Another structure operating under normal-to-plane propagation has also been proposed for magnetic resonance. It consists of a pair if aligned cut-wires or square plates, as shown in Figure 1.16(b) [DOL 05].

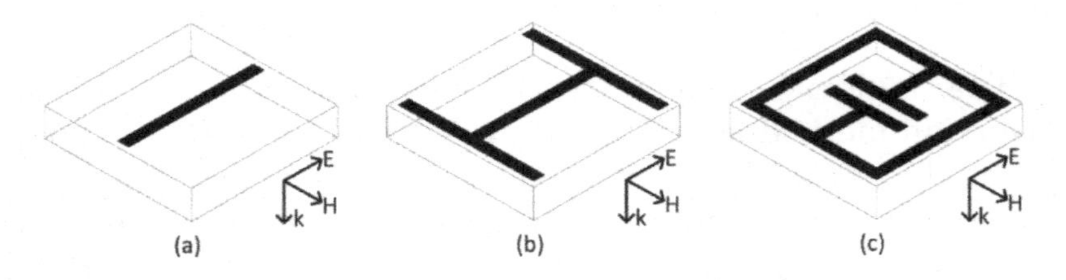

Figure 1.15. *Electric resonators: a) cut wire structure; b) I-shaped structure; c) ELC structure*

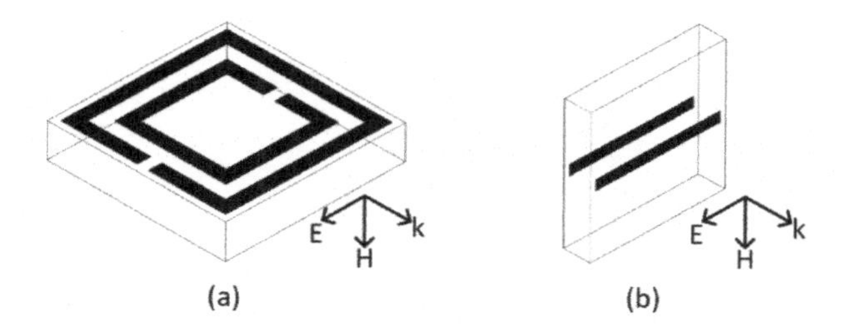

Figure 1.16. *Magnetic resonators: a) split ring resonator; b) cut-wire or plate pairs*

1.3.3. *All-dielectric material*

All-dielectric materials that will be considered to implement the TO devices detailed in this book are used to achieve engineered permittivity values. The relative permittivity value of a dielectric material can be modified and lowered by diluting it with air, i.e. by placing air holes in the materials. As such, making air holes with different radii in a dielectric material will allow it to produce a graded permittivity material.

However, it should also be noted that artificial magnetism is also possible by the enhancement of Mie resonances in high dielectric permittivity media [OBR 02].

1.4. Conclusion

This chapter is devoted to the presentation of the different aspect essential for transformation optics technique. Two types of transformations are detailed: coordinate transformation and space transformation. The different resonators currently used to engineer the material parameters required for the implementation of TO-based devices have also been presented.

2

Coordinate Transformation Concept: Transformation of Electromagnetic Sources

2.1. Introduction

In this chapter, we will deal with devices designed using the coordinate transformation concept. The technique is applied for the design of transformation media applied to microwave electromagnetic sources. Three devices are studied and designed. Theoretical formulations are presented to describe the transformations and full-wave simulations are performed to verify the functionality of the calculated devices. Anisotropic material parameters of the transformation-based devices are calculated and engineered using metamaterial elements such as split ring resonators and electric-LC resonators. The first device is conceived to transform a directive radiation pattern into an isotropic one through a space-stretching. An experimental validation of the device is also performed. The second device is a metamaterial shell capable of realizing a large aperture emission from a small aperture one through space compression. Such a device allows for miniaturizing electromagnetic sources considerably without altering the radiation properties. The third device is intended to create multiple beams from an isotropic emission. These designs illustrate the enormous possibilities provided by coordinate transformation for the transformation of electromagnetic emissions.

2.2. Isotropic antenna: transforming directive into isotropic pattern

In this section, we will describe how coordinate transformation can be applied to transform a directive emission into an isotropic one [TIC 11, TIC 13b].

It has long been supposed that isotropic radiation by a simple coherent source is impossible due to changes in polarization. Though hypothetical, the isotropic source is usually taken as a reference for determining an antenna's gain. Through coordinate transformation, we demonstrate, both theoretically and experimentally, that an isotropic radiator can be made up of a simple and finite source surrounded by a metamaterial shell. Such demonstration paves the way to other possibilities in optical illusion, wireless communication and antenna engineering.

2.2.1. *Theoretical formulations and numerical simulations*

The schematic principle illustrating the proposed method is presented in Figure 2.1. Consider a source radiating in a circular space as shown in Figure 2.1(a) where a circular region bounded by the black circle around this source limits the radiation zone. Here, a "space-stretching" technique is applied. The "space stretching" coordinate transformation consists of stretching exponentially the central zone of this delimited circular region represented by the shaded circle as illustrated in Figure 2.1(b).

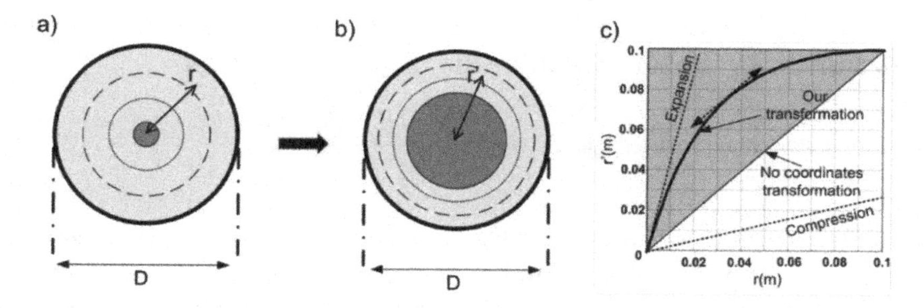

Figure 2.1. *Illustration showing the coordinate transformation for the isotropic antenna: a) initial space; b) transformed space; c) transformation rule made of a space stretching followed by a compression*

However, a good impedance matching must be kept between the stretched space and the exterior vacuum. Thus, the expansion procedure is followed by a compression of the annular region formed between the shaded and black circles so as to secure an impedance matching with free space. Figure 2.1(c) summarizes the exponential form of the coordinate transformation. The diameter of the transformed circular medium is denoted by *D*. Mathematically, the transformation is expressed as [TIC 11]:

$$\begin{cases} r' = \alpha(1 - e^{qr}) \\ \theta' = \theta \\ z' = z \end{cases} \quad \text{with } \alpha = \frac{D}{2} \frac{1}{1 - e^{\frac{qD}{2}}} \qquad [2.1]$$

where r', θ' and z' are the coordinates in the transformed cylindrical space, and r, θ and z are those in the initial cylindrical space. In the initial space, free space conditions are assumed with isotropic permittivity and permeability tensors ε_0 and μ_0. Parameter q (in m^{-1}) is an expansion factor, which can be viewed physically as the degree of space expansion. It must be negative to achieve the impedance matching condition. A high (negative) value of q means high expansion whereas a low (negative) value of q means nearly zero expansion. Calculations lead to permeability and permittivity tensors given in the diagonal base by:

$$\overline{\overline{\varepsilon}} = \overline{\overline{\psi}}\varepsilon_0 \text{ and } \overline{\overline{\mu}} = \overline{\overline{\psi}}\mu_0 \text{ with } \overline{\overline{\psi}} = \frac{A_i^{i'} A_j^{j'} \delta^{ij}}{\left|A_i^{i'}\right|} \qquad [2.2]$$

where δ^{ij} is the Kronecker symbol. We can observe that both electromagnetic parameters ε and μ share the same behavior, allowing for an impedance matching with that of the vacuum outside the transformed space. The inverse transformation is obtained from the initial transformation of equation [2.1] and is derived by a substitution method, enabling the circular metamaterial design which leads to anisotropic permittivity and permeability tensors.

By substituting the new coordinate system in the tensor components, and after some simplifications, the material parameters are derived. Calculations lead to permeability and permittivity tensors given in the diagonal base by:

$$\overline{\overline{\psi}} = \begin{pmatrix} \psi_{rr} & 0 & 0 \\ 0 & \psi_{\theta\theta} & 0 \\ 0 & 0 & \psi_{zz} \end{pmatrix} = \begin{pmatrix} \frac{qr(r'-\alpha)}{r'} & 0 & 0 \\ 0 & \frac{r'}{qr(r'-\alpha)} & 0 \\ 0 & 0 & \frac{r}{r'q(r'-\alpha)} \end{pmatrix} \qquad [2.3]$$

with $\mathrm{r} = \frac{\ln\left(1 - \frac{\mathrm{r}'}{\alpha}\right)}{q}$.

The components in the Cartesian coordinate system are calculated and are as follows:

$$\begin{cases} \psi_{xx} = \psi_{rr}\cos^2(\theta) + \psi_{\theta\theta}\sin^2(\theta) \\ \psi_{xy} = \psi_{yx} = (\psi_{rr} - \psi_{\theta\theta})\sin(\theta)\cos(\theta) \\ \psi_{yy} = \psi_{rr}\sin^2(\theta) + \psi_{\theta\theta}\cos^2(\theta) \end{cases} \qquad [2.4]$$

The ε and μ tensor components present the same behavior as given in equation [2.4]. Figure 2.2 shows the variation of the permittivity and permeability tensor components in the generated transformed space. The geometric dimension D is chosen to be 20 cm and parameter q is fixed to –40 m^{-1}. It can be noted that components ψ_{xx}, ψ_{yy} and ψ_{zz} present variations and an extremum that are simple to realize with commonly used metamaterials by reducing their inhomogeneous dependence. At the center of the transformed space, ε and μ present very low values (<< 1).

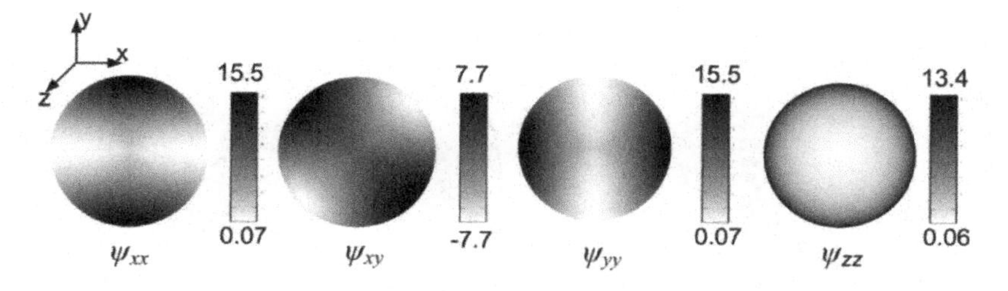

Figure 2.2. *Variation in the permeability and permittivity tensor components of the transformed space for D = 20 cm and q = –40 m^{-1}*

Consequently, light velocity and the corresponding wavelength in the transformed space are much higher than in a vacuum. The width of the source, then, appears very small with respect to wavelength and the source can, then, be regarded as a radiating wire, which is in fact an isotropic source. The specificity of this transformation indeed depends on the value of the expansion factor q and more generally, it can be applied to a wide range of electromagnetic objects, where the effective size can be reduced compared to a given wavelength.

Numerical verifications of the proposed concept using commercial solver COMSOL Multiphysics are performed to design and characterize the transformed quasi-isotropic emission. A plane current source is used as excitation in the simulations.

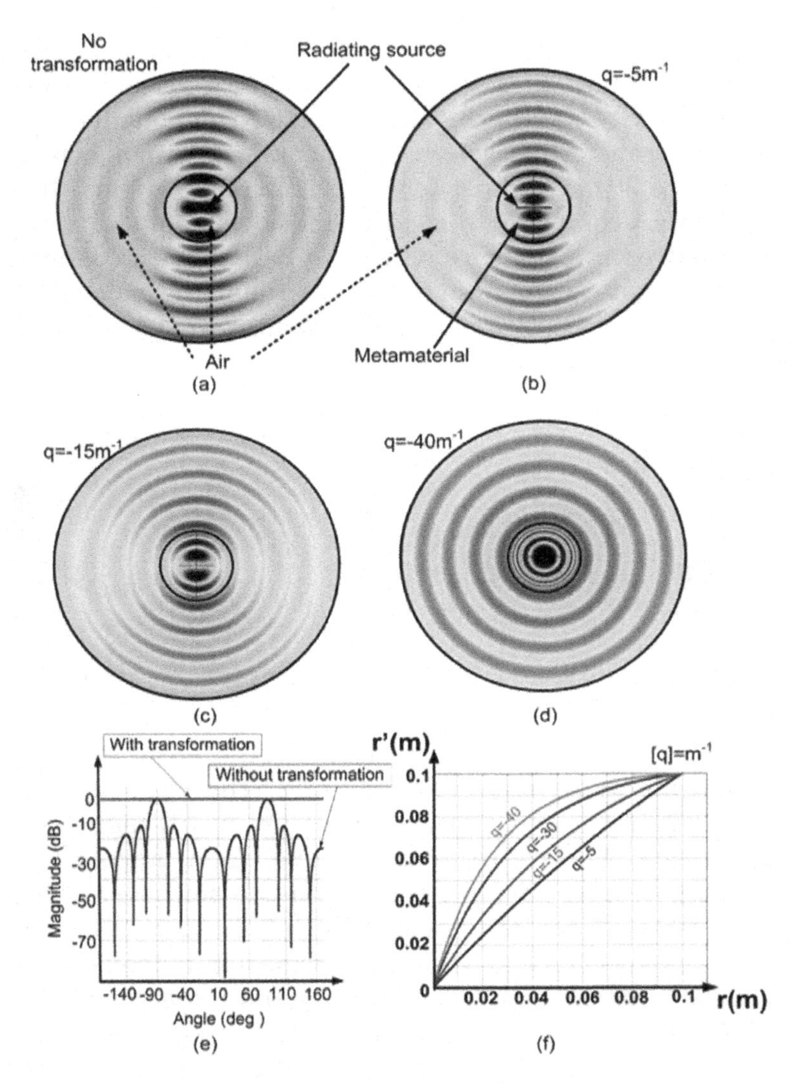

Figure 2.3. *Simulated electric field distribution for a TE wave polarization at 4 GHz [TIC 11]: a) a planar current source is used as the excitation for the transformation. The current direction is perpendicular to the plane of the figure; b)–d) verification of the transformation for different values of expansion factor q; e) 2D far field radiation pattern of the emission with (q = –40) and without transformation; f) influence of the parameter q on the proposed coordinate transformation. The emitted radiation is more and more isotropic as q tends to high negative values*

Figure 2.3 represents simulations results of the source radiating in the initial circular space at an operating frequency of 4 GHz. The current direction of the source is supposed to be along the z-axis. Simulations are performed in a Transverse Electric (TE) mode with the electric field polarized along z-direction. The surface current source is considered to have a width of 10 cm, which is greater than the 7.5 cm wavelength at 4 GHz. To plot the radiation properties, radiation boundary conditions are placed around the calculation domain.. Continuity of the field is assured in the interior boundaries.

As shown in Figure 2.3, different values for the expansion factor q are used in the simulations and as it can be observed, the emission becomes more isotropic gradually as the factor q decreases from -5 m^{-1} to -40 m^{-1}. As stated previously and verified from the different electric field distribution patterns, a high negative value of q leads to a quasi-perfect isotropic emission as the space expansion is higher. In Figure 2.3(e), we can note the far-field isotropic pattern of our device compared to the directive source. Figure 2.3(f) gives an insight of the influence of the parameter q on the proposed coordinate transformation.

2.2.2. *3D design and implementation using metamaterials*

The calculated material in equation [2.4] shows a dependence on the spatial coordinates, leading to an anisotropic material presenting non-diagonal terms. Thus, a reduction procedure and simplification of the materials parameters is highly desired. Choosing plane wave solutions for the electric field and magnetic field, with wave vector k in (r, θ) plane, it is possible to calculate the propagation equation in an inhomogeneous and anisotropic medium and obtain a dispersion relation which can be given by det(F) = 0, with:

$$F=\begin{pmatrix} \varepsilon_{rr}-\dfrac{k_\theta^2}{\mu_{zz}} & \dfrac{-k_r k_\theta}{\mu_{zz}} & 0 \\ \dfrac{-k_r k_\theta}{\mu_{zz}} & \varepsilon_{\theta\theta}-\dfrac{k_r^2}{\mu_{zz}} & 0 \\ 0 & 0 & \varepsilon_{zz}-\dfrac{k_r^2}{\mu_{\theta\theta}}-\dfrac{k_\theta^2}{\mu_{rr}} \end{pmatrix} \qquad [2.5]$$

Taking into account the fact that our source is a printed electric dipole antenna that delivers a bi-directional emission in the E-plane, we favor a TM polarization of the electromagnetic field so that we suppose having the magnetic field along the z-direction. Solving det(F) = 0, the equation for TM polarization is written as

$$\mu_{zz}=\frac{k_r^2}{\varepsilon_{\theta\theta}}+\frac{k_\theta^2}{\varepsilon_{rr}} \qquad [2.6]$$

In this case the relevant electromagnetic parameters are μ_{zz}, ε_{rr} and $\varepsilon_{\theta\theta}$. The associated dispersion equation can, then, be modified by dividing equation [2.6] by μ_{zz}, without changing propagation in the structure. The following reduced material parameters can then be obtained:

$$\begin{cases} \mu_{zz}=1 \\ \varepsilon_{rr}=\left(\dfrac{r}{r'}\right)^2 \\ \varepsilon_{\theta\theta}=\left(\dfrac{1}{q(r'-\alpha)}\right)^2 \end{cases} \qquad [2.7]$$

The main drawback of using such parameters is the impedance mismatch at the boundary with vacuum. Indeed, modifying the dispersion equation renders permittivity and permeability non-equal ($\varepsilon \neq \mu$). So, a possible approximation is to change the value of α without changing the dimensions of the structure and the expansion factor q. In such case, this can be viewed as a translation of our transformation in a topological point of view. As explained previously, the transformation is made of a space-stretching and a space-compression to assure perfect impedance matching. Since, the reduction of parameters creates an impedance mismatch with vacuum, we can, therefore, truncate the material before compression of space is performed in the perfect transformation.

The final reduced parameters are, thus, simplified and illustrated by the continuous traces plotted in Figure 2.4. μ_{zz} is constant along z and equal to 1. The radial permittivity $\varepsilon_{\theta\theta}$ has a maximum value around 6 and the angular permittivity ε_{rr} varies from 0.05 to 0.07 for a radius varying from 0 to 5 cm.

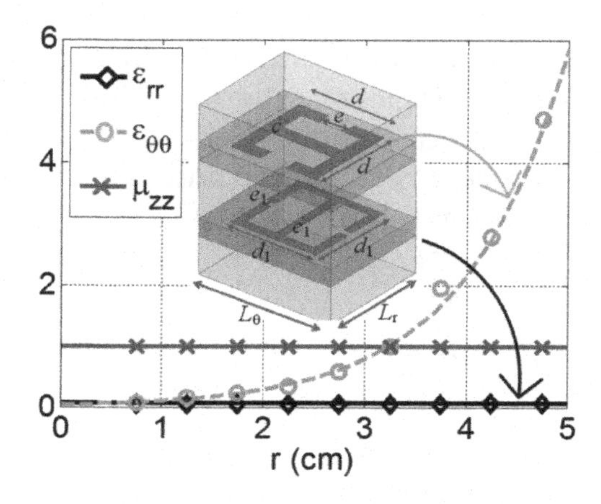

Figure 2.4. *Final reduced material parameters for a perfect impedance matching. The inset shows the building block composed of two ELC resonators used to provide ε_{rr} and $\varepsilon_{\theta\theta}$*

The parameters present positive values which can be easily achieved firstly by discretizing their continuous profile and secondly by using electric-LC (ELC) resonators where electric responses can be tailored and controlled.

As presented by the inset in Figure 2.4, the building block is composed of two ELC resonators; one responding to the angular component and the other one to the radial component of the electric field. The implemented metamaterial unit cell is not periodic, and two types of elementary resonators are optimized at one go, since common parameters are shared. A microstrip planar dipole antenna is used as feed in order to provide both the angular and radial components of the electric field. The dielectric substrate used for the printed circuit boards is the 0.787 mm thick low loss ($\tan\delta = 0.0013$) Rogers RT/Duroid 5870™, with relative permittivity $\varepsilon_r = 2.33$.

The surrounding material, made up of alternating layers of electric angular and radial metamaterial permittivity, transforms the directive emission in the xy plane of the dipole into an isotropic one. Choosing a polarized electromagnetic field along the z-direction allows to fix $\mu_{zz} = 1$ and control $\varepsilon_{\theta\theta}$ and ε_{rr} only. The material is composed of nine different regions where radial and angular permittivities vary accordingly. Considering constraints of the layout, we chose a rectangular unit cell with dimensions $L_r = 5$ mm except for the first zone ($L_r = 20/3$ mm chosen). We also chose $L_\theta = \frac{\pi}{NL_r}\left[R_n^{\,2} - R_{n-1}^{\,2}\right]$, where N represents the number of patterns in the nth zone and R_n, the radius of the nth zone.

The scattering matrix S is calculated using a commercial electromagnetic solver. As stated in Chapter 1, the material parameters can then be extracted by using an inversion method. The different geometrical dimensions used in the metamaterial building blocks and the permittivity values extracted for the nine zones constituting the metamaterial layers are summarized in Table 2.1.

Zone	L_θ (mm)	L_r (mm)	c (mm)	d (mm)	e (mm)	c_1 (mm)	d_1 (mm)	e_1 (mm)	$\varepsilon_{\theta\theta}$	ε_{rr}
1	7	6.666	0.325	4	1.85	0.308	4	1.85	0.07	0.05
2	6.5	5	0.385	4	1.85	0.479	4	1.85	0.15	0.04
3	6.9	5	0.3	4	1.77	0.468	4	1.85	0.23	0.04
4	6.4	5	0.35	4	1.85	0.5	4	1.85	0.35	0.05
5	5.4	5	0.3	4	1.8	0.425	4	1.85	0.58	0.05
6	5.1	5	0.3	4.1	1.9	0.48	4	1.85	1	0.085
7	5.9	5	0.9	3	1	0.3	4	1.85	1.96	0.045
8	5.6	5	0.83	3.8	1.3	0.365	4	1.85	2.8	0.07
9	6.2	5	0.59	4	1.4	0.4	4	1.85	4.71	0.07

Table 2.1. *Geometrical dimensions and material parameters extracted*

The discrete set of simulations and extractions was interpolated to obtain particular values of the geometric parameters that yielded the desired material properties plotted in Figure 2.4. As it can be observed, ε_{rr} takes values close to zero, a consequence of the expansion factor $q = -50$ so as to achieve quasi-perfect isotropic emission in the xy plane. The impedance matching is assured in the proposed implementation.

2.2.3. *Experimental validation of fabricated isotropic antenna*

Figure 2.5 shows a photograph of the fabricated prototype, illustrating the angular and radial permittivity gradient layers and the feed source used. The bulk metamaterial is assembled using 30 layers of the dielectric substrates on which the subwavelength resonant ELC structures are printed. The layers are mounted 1 × 1, alternating angular and radial permittivity layers, with a constant spacing of 3.3 mm between each layer. Overall dimensions of the antenna are D × H = 10 cm × 12 cm, where D is the diameter and H, the height.

To validate the isotropic antenna performances, electric near field is measured in the different planes shown in Figure 2.6.

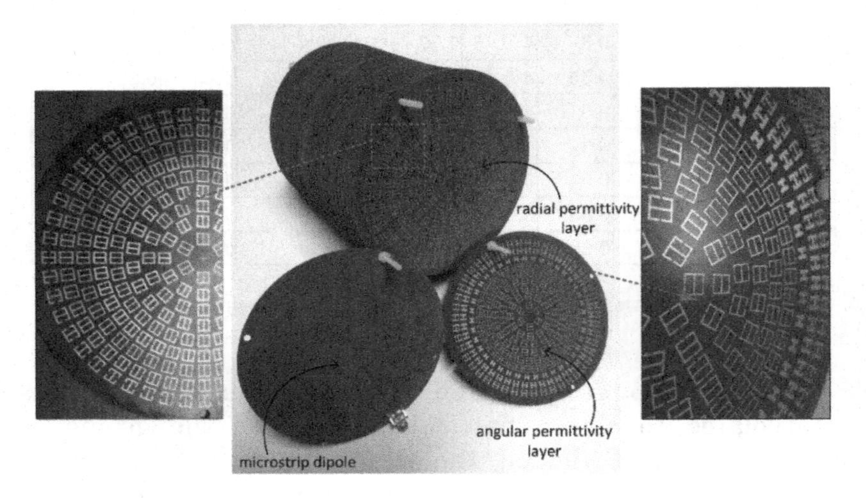

Figure 2.5. *Photograph of the fabricated structure using alternating layers of angular and radial permittivity gradient layers. Each layer is discretized into 9 zones providing the material parameters profile necessary for the coordinate transformation [TIC 13b]*

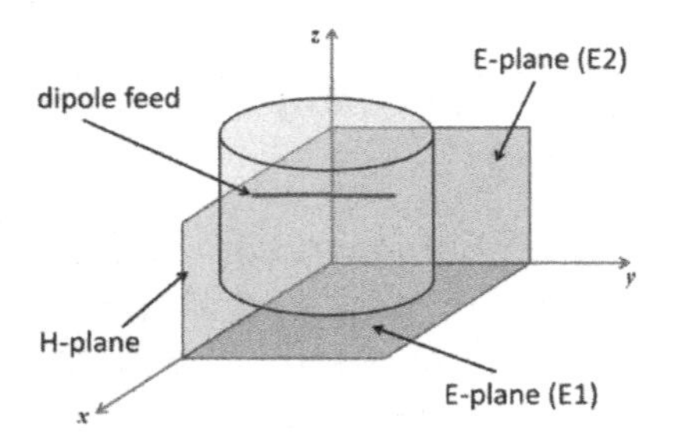

Figure 2.6. *Measurement planes used to map the near-fields*

The E-field is scanned by a field-sensing monopole probe connected to the network analyzer by a coaxial cable. The probe is mounted on two orthogonal linear translation stages (computer-controlled Newport MM4006), so that the probe can be translated with respect to the radiation region of the antenna. By stepping the field sensor in small increments and recording the field amplitude and phase at every step, a full 2D spatial field map of the microwave near field pattern is acquired in the free-space radiation region. The total scanning area covers 400 × 400 mm^2 with a step resolution of 2 mm in lateral dimensions. Microwave absorbers are applied around the measurement stage to suppress undesired scattered radiations at the boundaries.

Figure 2.7 shows the near-field mappings for both the dipole feed and the metamaterial-based antenna. As can be observed along the right-hand side, the transformed medium designed with ELC resonators allows for producing an omnidirectional mapping in both the E-planes. In the H-plane, the radiation of the dipole alone is intrinsically omnidirectional. Hence, no substantial difference is observed in the field mappings with the transformed medium in the H-plane. Therefore, the antenna presents omnidirectional patterns in all three directions and such an example shows the potentials of metamaterials combined with transformation optics technique to realized innovative concepts.

2.3. Miniaturization of electromagnetic sources

In this section, coordinate transformation is used to tailor the electromagnetic appearance of radiating elements. We create a large aperture emission from a small aperture one. In the telecommunications domain, there are actually growing interests in the miniaturization of devices, particularly for antennas in transport and aeronautical fields. In most cases, it is the physics itself that limits the possibility of size reduction.

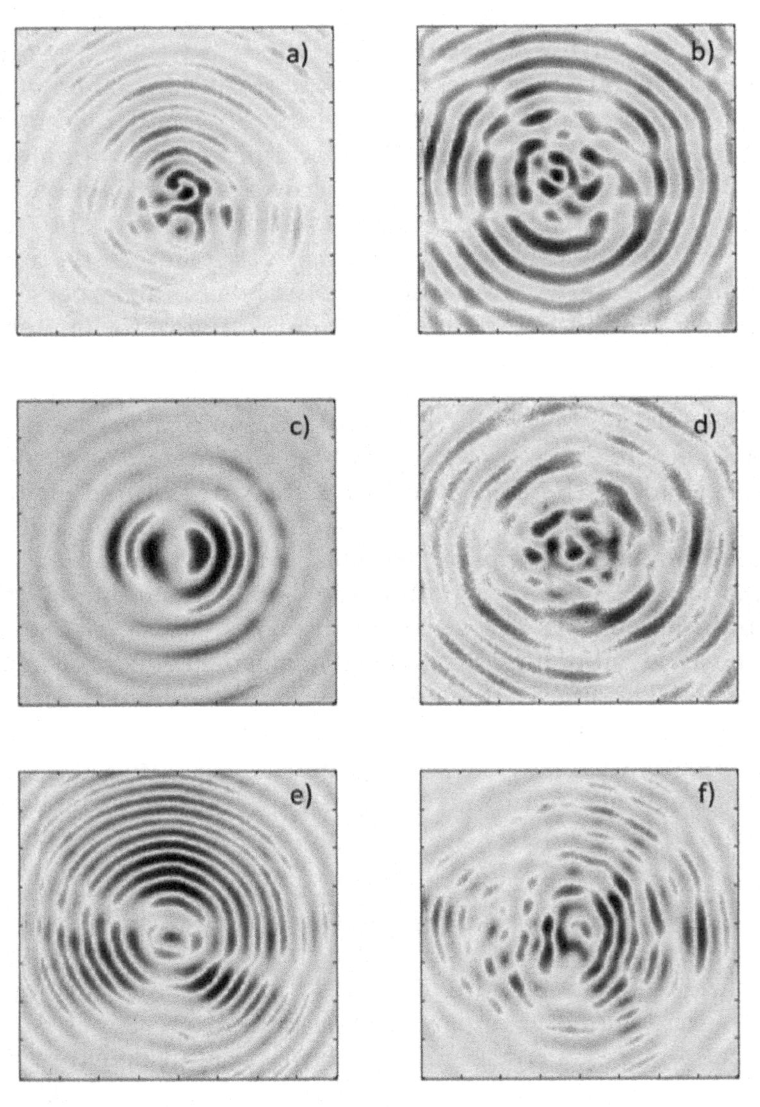

Figure 2.7. *Measured near-field mappings for dipole feed (left column) and metamaterial antenna (right column): a)–b) E-plane (E1); c)–d) E-plane (E2); e)–f) H-plane*

Through coordinate transformation, we demonstrate theoretically that directive antenna can be miniaturized without altering its radiation pattern. We study a transformed medium that is able to make a very small radiator produce an electromagnetic emission as if it were several times bigger. The transformation allows us to have a directive emission from a miniaturized (dimensions much smaller than the operating wavelength) radiating element, which, in general, radiates in an isotropic manner.

2.3.1. *Theoretical formulations*

To achieve the transformation of a small aperture source into a much larger one, we have to discretize the space around the latter radiating element into two different zones; a first zone which will make our source appear bigger than its real physical size and a second zone which ensures an impedance matching with the surrounding radiation environment. The transformation principle is shown by the schematic in Figure 2.8(a) [TIC 13a]. Therefore, the technique consists of compressing a circular region of space of radius R_1/q_1 (with $q_1 < 1$), represented by the shaded circle in Figure 2.8(a) in a region of radius R_1. In the studied transformation, the space is described by polar coordinates and the angular part of these coordinates remains unchanged. The second part of the transformation consists of an impedance matching with the surrounding space through an annular expansion zone defined between circular regions with radius R_1 and R_2, as illustrated in Figure 2.8(b). This space expansion can be performed using three different types of transformations: a positive exponential transformation, a negative exponential transformation, and a linear one. We denote below, and in the rest of the chapter, the two different zones by the index i, where $i = 1$ corresponds to the first zone (compression) and $i = 2$ corresponds to the second zone (expansion). The final virtual space describing our device is represented in Figure 2.8(c).

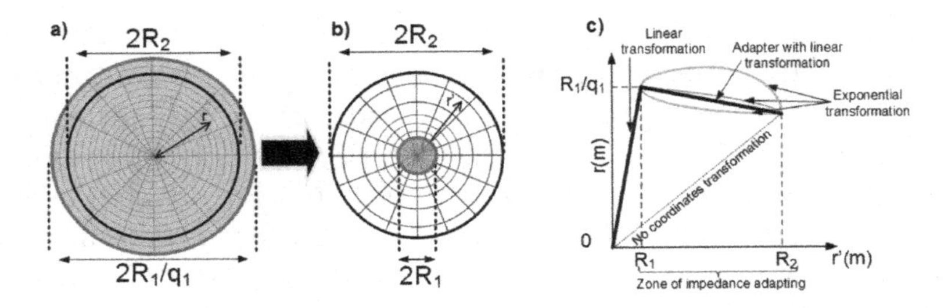

Figure 2.8. *Representation of the proposed coordinate transformation: a) initial and b) virtual space; c) operating principle of the transformation, which consists firstly of a compression of the central space ($0 < r' < R_1$) and secondly, an expansion to match the space metric ($R_1 < r' < R_2$). Continuity of the transformations is assured at the boundary of the compressed region (point A) and at the outer boundary of the device (point B) [TIC 13a]*

Figure 2.8(d) summarizes the different transformations considered. To secure the impedance and metric matching, continuity of the transformations is assured at the boundary of the first region and at the outer boundary of the device, respectively noted point A and point B in Figure 2.8(d).

Mathematically, the general transformation in the two regions is written as:

$$\begin{cases} r' = f_i(r,\theta) \\ \theta' = \theta \\ z' = z \end{cases} \qquad [2.8]$$

The Jacobian matrix of the transformation is given in the cylindrical coordinate system as:

$$A_{cyl} = \begin{pmatrix} \dfrac{\partial r'}{\partial r} & \dfrac{\partial r'}{\partial \theta} & \dfrac{\partial r'}{\partial z} \\ \dfrac{\partial \theta'}{\partial r} & \dfrac{\partial \theta'}{\partial \theta} & \dfrac{\partial \theta'}{\partial z} \\ \dfrac{\partial z'}{\partial r} & \dfrac{\partial z'}{\partial \theta} & \dfrac{\partial z'}{\partial z} \end{pmatrix} = \begin{pmatrix} f_{i,r} & f_{i,\theta} & 0 \\ 0 & 1 & 0 \\ 0 & 0 & 1 \end{pmatrix} \qquad [2.9]$$

where $f_{i,r}$ and $f_{i,\theta}$ represent the respective derivatives of f_i with respect to r and θ. To calculate permittivity and permeability tensors directly from the coordinate transformation in the cylindrical and orthogonal coordinates, we need to express the metric tensor in the initial and virtual spaces. The final Jacobian matrix needed for the permeability and permittivity tensors of our material is then given as:

$$A_i^{i'} = \begin{pmatrix} f_{i,r} & \dfrac{f_{i,\theta}}{r} & 0 \\ 0 & \dfrac{r'}{r} & 0 \\ 0 & 0 & 1 \end{pmatrix} \qquad [2.10]$$

The material parameters obtained using the transformation in equation [2.8] are:

$$\begin{cases} (\psi_{rr})_i = \dfrac{rf_{i,r}}{r'} + \dfrac{f_{i,\theta}^2}{rr'f_{i,r}} \\ (\psi_{r\theta})_i = \dfrac{f_{i,\theta}}{rf_{i,r}} \\ (\psi_{\theta\theta})_i = \dfrac{r'f_{i,r}}{r} \\ (\psi_{zz})_i = \dfrac{r}{r'f_{i,r}} \end{cases} \qquad [2.11]$$

These parameters are relatively simple for the transformation in the first zone since it leads to constant values. But the permittivity and permeability components have to be expressed in the Cartesian coordinate system so as to have a perfect equivalence in Maxwell's equations and also to design our device physically. Using matrix relations between cylindrical and Cartesian coordinates, we have:

$$\overline{\overline{\varepsilon}} = \begin{pmatrix} \psi_{xx} & \psi_{xy} & 0 \\ \psi_{yx} & \psi_{yy} & 0 \\ 0 & 0 & \psi_{zz} \end{pmatrix} \varepsilon_0 \text{ and } \overline{\overline{\mu}} = \begin{pmatrix} \psi_{xx} & \psi_{xy} & 0 \\ \psi_{yx} & \psi_{yy} & 0 \\ 0 & 0 & \psi_{zz} \end{pmatrix} \mu_0 \quad [2.12]$$

with

$$\begin{cases} \psi_{xx} = \psi_{rr} \cos^2(\theta) + \psi_{\theta\theta} \sin^2(\theta) - \psi_{r\theta} \sin(2\theta) \\ \psi_{xy} = \psi_{yx} = (\psi_{rr} - \psi_{\theta\theta}) \sin(\theta) \cos(\theta) + \psi_{r\theta} \cos(2\theta) \\ \psi_{yy} = \psi_{rr} \sin^2(\theta) + \psi_{\theta\theta} \cos^2(\theta) + \psi_{r\theta} \sin(2\theta) \end{cases} \quad [2.13]$$

The angular part in the coordinate transformation described above allows for obtaining more general and adjustable parameters for a possible physical realization of the device. However, we consider $f_{i,\theta} = 0$ here to simplify the calculations. To apply the proposed coordinate transformation, we consider a radial compression of the space in region 1. This leads to a material with high permittivity and permeability tensors. For the transformation, we choose

$$r' = f_1(r) = q_1 r \quad [2.14]$$

with q_1 being the compression factor applied in the central region and lower than 1. This factor has a transition value, which can be defined as $q_c = R_1/R_2$ where the material of the matching zone (region 2 in Figure 2.8(c)) switches from a right-handed (positive refractive index) to a left-handed (negative refractive index) material. Indeed, when $q_1 < q_c$ the material presents a negative index and the final apparent size of the source can be larger than $2R_2$. Now, if this embedded source has a small aperture, much smaller than the wavelength, then after transformation this antenna will behave like one with a large aperture, typically q_1 times larger and potentially much greater than the wavelength. A small aperture antenna is well known to radiate in an isotropic manner. The same antenna embedded in the material defined by equation [2.13] will present a directive radiation, and therefore, appear electrically as if its size is larger than the working wavelength. Moreover, we can obtain the radiation of a conventional array of antennas using much smaller dimensions for the latter array embedded in zone 1.

To ensure a good impedance matching for the radiated fields, a matching zone (region 2) is added around region 1. To design this zone, three different possible transformations can be considered to match the space from R_1 to R_2. The first studied transformation for this matching region is a linear one that takes the form

$$r' = f_2(r) = \frac{1}{\alpha}\left[r + R_2(\alpha - 1)\right] \qquad [2.15]$$

where as the two other transformations are of logarithmic forms that can be expressed as

$$r' = f_2(r) = \frac{1}{q_2}\ln\left(\frac{r-d}{p}\right) \qquad [2.16]$$

where $d = R_2 - pe^{q_2 R_2}$ and $p = \dfrac{R_2 - q_1 R_1}{e^{q_2 R_2} - e^{q_1 R_1}}$ are constant values. In these two cases, the inverse transformation defining r from r' has an exponential form defined by $F_2(r') = d + pe^{q_2 r}$. This exponential transformation can be characterized by the factor q_2 that indicates the shape of the progressive metric matching to vacuum, as illustrated in Figure 2.8(d). A small value of q_1 indicates a high compression of the space in the first region. To compensate this high compression, the transformation in region 2 gives negative electromagnetic parameters due to the relative positions of points A and B, as presented by the components of the permittivity and permeability tensors in Figure 2.9. In such a case, the wave propagates with a backward phase in this region. For the linear transformation, the minimum and maximum of the material parameters depend on the geometrical properties of the problem and, thus, they depend only on α and γ which is given by:

$$\alpha = \frac{R_2 - \frac{R_1}{q_1}}{R_2 - R_1} \text{ and } \gamma = 1 + \frac{R_2(1-\alpha)}{R_1\,\alpha} \qquad [2.17]$$

where q_1 is defined on]0, 1]. α is, therefore, defined on]-∞,1] and vanishes at $q_1 = q_c$. Thus, γ is a function of α and is larger than 1 for $q_1 > q_c$ and is negative for $q_1 < q_c$. In the last case, such a medium is a left-handed material.

The trends of permittivity and permeability values in the Cartesian coordinates are quite similar for both linear and exponential transformation. The values depend only on R_1, R_2, q_1 and q_2 for the linear transformation. For the case of the exponential transformation in region 2, the parameters considered are $q_1 = 1/16$, $q_2 = 15$, $R_1 = 5$ mm and $R_2 = 45$ mm and as it can be observed, the calculated components ψ_{xx}, ψ_{yy} and ψ_{zz} are always negative.

2.3.2. *Numerical simulations*

Calculations are performed in a two-dimensional configuration with a transverse electric mode (TE_z) (E parallel to the z-axis) to verify the functionality of the device at 10 GHz. Different current sources perpendicular to the xy plane are used as radiating elements to show that the transformation can be applied to any type of source embedded in the region 1. Continuity and matched conditions are applied respectively to the boundary of zone 1 and zone 2.

We fix R_1 = 2 mm and R_2 = 45 mm. The results obtained from linear transformations both in region 1 and 2, as defined by the continuous black trace in Figure 2.9(a), are presented in Figure 2.10. In Figure 2.10(a), the electric field distribution of a current source radiating in free space is plotted. The source is supposed to have a width d = 80 mm (2.7λ at 10 GHz). For such a large size, the radiation is equivalent to that of an array of several elements and, therefore, the radiated field is directive. Figure 2.10(b) shows a similar source but with a much smaller size d = 2 mm (λ/15 at 10 GHz) embedded in the metamaterial shell having a compression factor q_1 = 1/40. In this scenario, a radiation pattern similar to the large aperture source is observed, demonstrating that small aperture antennas inserted in the proposed material shell present the same electromagnetic behavior as much larger aperture antennas in free space. This same miniature source will, however, radiate in an isotropic manner in free space (Figure 2.10(c)). The same observations can be made when replacing the linear current source by a crossed-type one, as illustrated in Figure 2.10(d)–(f).

Furthermore, the transformation holds for a phased-array antenna. We consider an array of three sources of length L = 12.5 mm, spaced by a distance a = 5 mm and with a 30° phase shift between each element. These sources radiate in vacuum with a beam pointing in an off-normal direction due to the phase shift applied between the different elements of the array. When the dimensions of these antennas are reduced by a factor of 25 (q_1 = 1/25), the dimensions of the array become smaller compared to the wavelength and the radiation pattern in free space is isotropic as shown in Figure 2.11(b). By embedding the small sources in a material defined from the double linear transformation, we are able to recover the beam steering of the source array as shown in Figure 2.11(c).

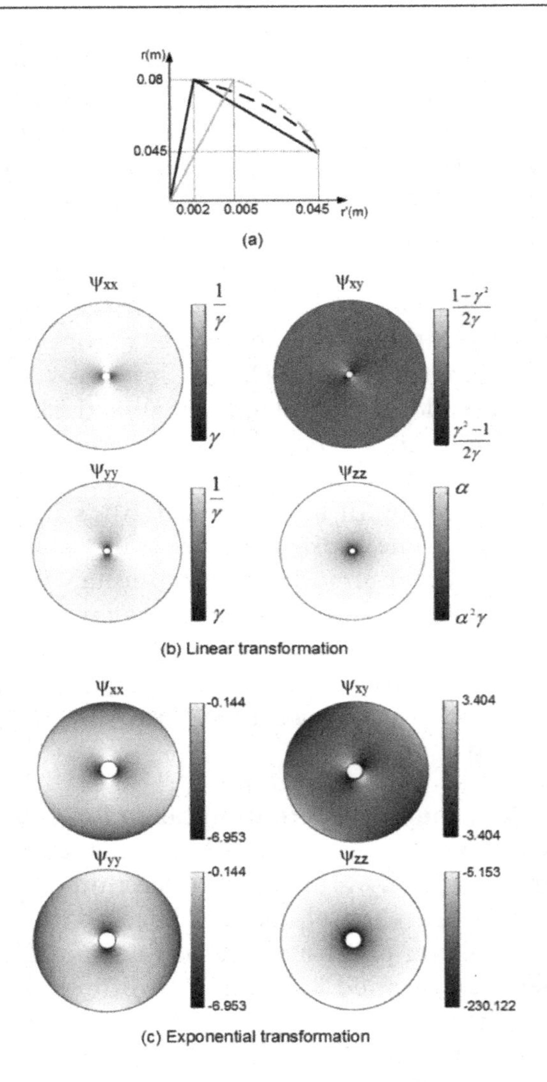

Figure 2.9. *a) Representation of the transformations in regions 1 and 2. The black and gray traces correspond respectively to q_1 = 1/40 and q_1 = 1/16 and the continuous and dashed traces correspond respectively to a linear and exponential transformations; b)–c) variation of the components in Cartesian coordinates of the matching region 2. The permittivity and permeability are respectively plotted for the linear transformation with q_1 = 1/40, R_1 = 2 mm (continuous black traces case) and for the exponential transformation with q_1 = 1/16, q_2 = 15 and R_1 = 5 mm (continuous and dashed gray traces case) [TIC 13a]*

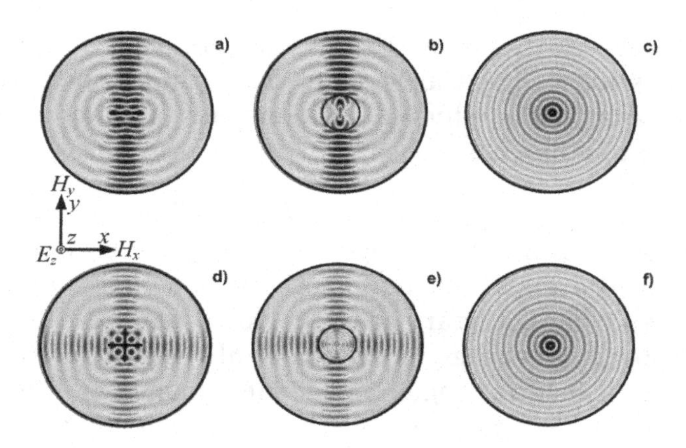

Figure 2.10. *Electric field (E_z) distribution at 10 GHz of a linear source: a) with dimension d = 80 mm radiating in free space, b) with dimension d = 2 mm embedded in a metamaterial shell and c) with dimension d = 2 mm radiating in free space. Electric field distribution at 10 GHz of a crossed-type source: d) with dimension d = 80 mm radiating in free space, e) with dimension d = 2 mm embedded in a metamaterial shell and f) with dimension d = 2 mm radiating in free space. The metamaterial shell is defined by a double linear transformation where R_1 = 2 mm, R_2 = 45 mm and q_1 = 1/40 [TIC 13a]*

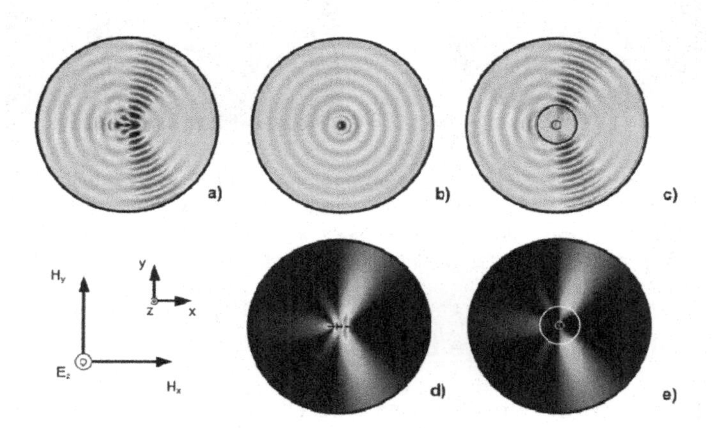

Figure 2.11. *Electric field (E_z) distribution at 10 GHz of 3 sources with 30° phase shift between each element: a) with length L = 12.5 mm spaced by a distance a = 5 mm radiating in free space, b) with L = 0.5 mm and a = 0.2 mm (all dimensions are reduced by a factor of 25) radiating in free space, c) with L = 0.5 mm and a = 0.2 mm embedded in the metamaterial shell defined by a double linear transformation. The metamaterial shell is defined by a double linear transformation with q_1 = 1/25 [TIC 13a]*

The above results show that we are, indeed, able to hide the physical appearance of radiating sources by miniaturizing their physical dimensions without altering their radiation diagrams.

It is also very important to stress the importance of the space expansion around the compressed space region. In the absence of the matching region, there is a high impedance mismatch at the boundary of the region 1 and all the energy emitted by the source is reflected at the boundary and confined in this latter region. This phenomenon is illustrated in Figure 2.12 by the norm of the electric field. Stationary waves appear in the structure due to reflection at $r = R_1$.

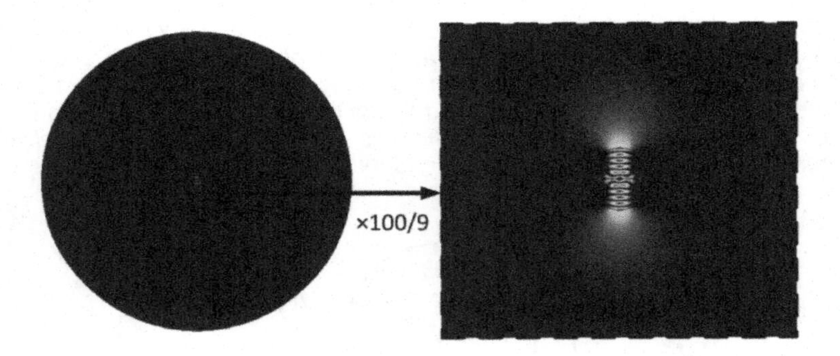

Figure 2.12. *Norm of the electric field of a line source with dimension d = 2 mm. The metamaterial shell is defined by only a compression region presenting a linear transformation with q_1 = 1/40. No matching region is used in this case [TIC 13a]*

2.4. Creation of multiple beams

In this section, we will describe the design of a device able to transform an omnidirectional beam into several directive beams. Furthermore, we will show that adjusting the transformation enables to control the number and the angular direction of the beams radiated.

2.4.1. *Theoretical formulations*

We consider a radiating source with an aperture much smaller than the wavelength, therefore, isotropic in the xy plane. To achieve the transformation of this small aperture source into several beams that are steered, we discretize the space around the latter radiating element into a region which will split the isotropic radiation into steered directive beams. The operating principle is shown by the schematic in Figure 2.13 [TIC 14]. In the zone defined between circular regions with radius R_1 and R_2, we apply a transformation of the angular part in function of the radial part which assures perfect impedance matching at some specific locations on the external material boundary (Figure 2.13(b)). Figure 2.13(c) presents the angular part transformation used, which can be performed using three different transformations: a positive exponential transformation, a negative exponential transformation or a linear one. A free parameter q allows adjusting electromagnetic achievable parameters of the metamaterials for realization.

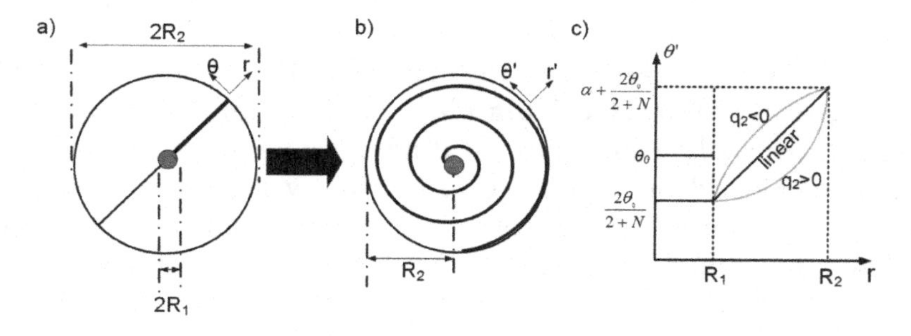

Figure 2.13. *Representation of the proposed coordinate transformation: a) initial space and b) virtual space. The device is composed of a region that allows creating several radiation beams and rotating them in a spiral way as indicated by the transformation of the thick black line; c) the θ-dependent transformation which can be linear or exponential (positive or negative) [TIC 14]*

In a space point of view, the virtual space, which assures the properties of our material, has been designed such that the angular lines are curved in function of the position in the space. The boundary material is fixed and has a radius R_2 represented by the external circle.

Mathematically, the transformation can be written as:

$$\begin{cases} r' = r \\ \theta' = f(r,\theta) \\ z' = z \end{cases} \qquad [2.18]$$

The Jacobian matrix of the transformation is given in the cylindrical coordinate system as:

$$A_{cyl} = \begin{pmatrix} \frac{\partial r'}{\partial r} & \frac{\partial r'}{\partial \theta} & \frac{\partial r'}{\partial z} \\ \frac{\partial \theta'}{\partial r} & \frac{\partial \theta'}{\partial \theta} & \frac{\partial \theta'}{\partial z} \\ \frac{\partial z'}{\partial r} & \frac{\partial z'}{\partial \theta} & \frac{\partial z'}{\partial z} \end{pmatrix} = \begin{pmatrix} 1 & 0 & 0 \\ f_r & f_\theta & 0 \\ 0 & 0 & 1 \end{pmatrix} \qquad [2.19]$$

where f_r and f_θ represent the respective derivatives of f with respect to r and θ. To calculate permittivity and permeability tensors directly from the coordinate transformation in the cylindrical and orthogonal coordinates, we need to express the metric tensor in the initial and virtual spaces. The final Jacobian matrix needed for the permeability and permittivity tensors of our material is, then, given as:

$$A_i^{i'} = \begin{pmatrix} 1 & 0 & 0 \\ r' f_r & \frac{r'}{r} f_\theta & 0 \\ 0 & 0 & 1 \end{pmatrix} \qquad [2.20]$$

The material parameters, obtained using the transformation in equation [2.18] are:

$$\begin{cases} \psi_{rr} = \dfrac{r}{r' f_\theta} \\ \psi_{r\theta} = \dfrac{r f_r}{f_\theta} \\ \psi_{\theta\theta} = \dfrac{r r' f_r^2 + \dfrac{r'}{r} f_\theta^2}{f_\theta} \\ \psi_{zz} = \dfrac{r}{r' f_\theta} \end{cases} \qquad [2.21]$$

The material parameters in Cartesian coordinates are the same as in equations [2.12] and [2.13].

To create multiple steered beams, a transformation on the angular part is performed. This transformation can be of two different types so as to divide the space into different segments and to steer the beam in each segment. First, we can consider a linear transformation that takes the form:

$$\theta' = f(r,\theta) = \frac{\theta}{1+\dfrac{N}{2}} + \alpha\left(\frac{r-R_1}{R_2-R_1}\right) \qquad [2.22]$$

The second type of transformation can be of exponential form that is expressed as:

$$\theta' = f(r,\theta) = \frac{\theta}{1+\dfrac{N}{2}} + \alpha\left(\frac{e^{q(r-R_1)}-1}{e^{q(R_2-R_1)}-1}\right) \qquad [2.23]$$

The transformations used are general and to validate the possibility of these transformations, we need to consider different cases. The denominator $1 + N/2$ of the function θ' has to be an integer where, N, can only take 0 or positive even values. For example, when N = 0, there is no multi-beam creation due to non-segmentation of the space.

Figure 2.14 shows the variation of the tensor for the exponential transformation case and when N is different from zero, so as to be in a multi-beam configuration. The parameters are q = –30, N = 2, α = 300°. We also fix R_1 = 5 mm and R_2 = 50 mm.

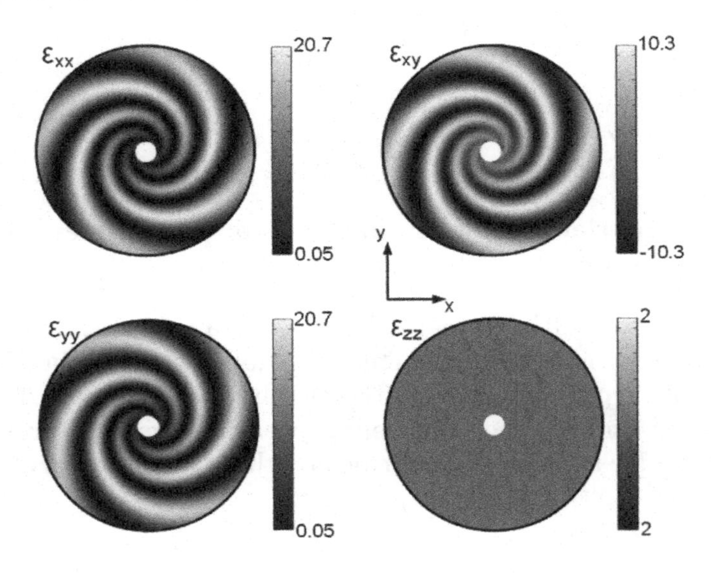

Figure 2.14. *Variation of the components in Cartesian coordinates. The permittivity and permeability are respectively plotted for the exponential transformation with* q = –30, R_1 = 5 mm, R_2 = 50 mm, N = 2 *and* α = 300°

2.4.2. *Numerical validation*

In Figure 2.15, the linear transformation of the radial part is followed by an exponential one with N different from zero, q = –30 and α = 300°, corresponding to the material parameters presented in Figure 2.14. Figures 2.15(a) and (b) show the electric field distribution in the proposed device for N = 2. Two steered beams can be clearly observed. The cases for N = 4 and N = 6 are shown in Figures 2.15(e) and (f) and

Figures 2.15(g) and (h), respectively. In each case, the electromagnetic field is rotated in the material as shown in Figures 2.15(c) and (d). When α increases, the radiation tends to be more and more tangential to the surface of the material, and the interferences observed in Figures 2.15(g) between the emitted beams decrease in Figures 2.15(h).

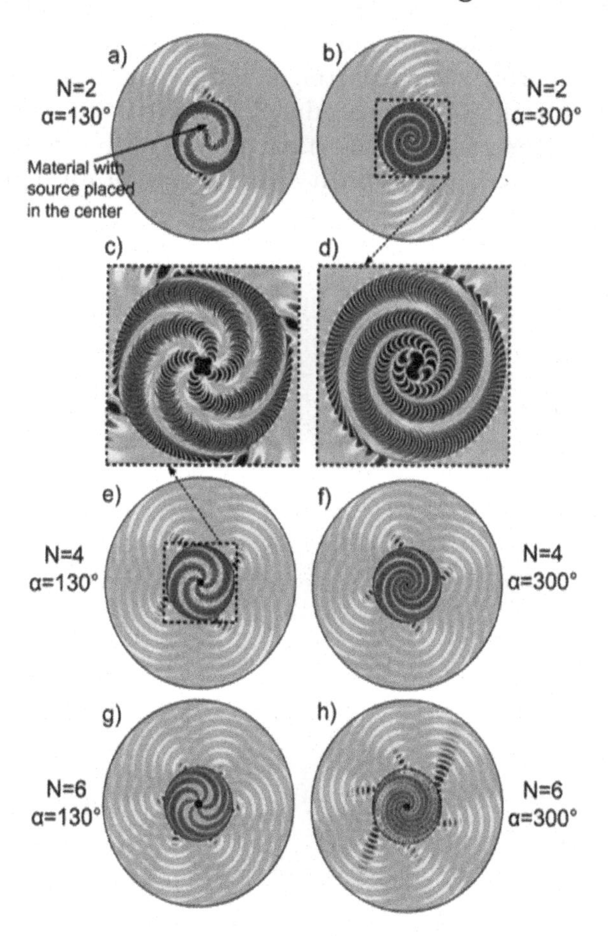

Figure 2.15. *Electric field (z-component) distribution of a source with dimension d = 4 mm in a material defined with q = -30, α = 130° (a, c, e, g) and 300° (b, d, f, h) at 20 GHz. Multi-beam emission is shown for N = 2 (a, b), N = 4 (e, f) and N = 6 (g, h). The electromagnetic field is rotated in the material as shown in the zooms presented in parts (c) and (d)*

2.5. Conclusion

This chapter is devoted to the transformation of electromagnetic sources. Coordinate transformations are performed to modify the antenna's radiation patterns. We showed how to produce an isotropic pattern from a directive one by stretching the space in which the source is embedded. We also show how transformation optics can be applied to restore the radiation patterns of antennas after their miniaturization. This procedure implies transforming an isotropic pattern into a directive one. Finally, we also focus our attention on the possibility of creating of multiple beams by space segmentation.

3

Space Transformation Concept: Controlling the Path of Electromagnetic Waves

3.1. Introduction

In this chapter, we will deal with devices designed using space transformation concept. Quasi-conformal transformation optics (QCTO) is applied for the design of electromagnetic devices at microwave frequencies. Antenna applications are targeted for focusing and collimating properties. Two lenses are studied and designed by solving the Laplace's equation that describes the deformation of a medium in the space transformation. The material parameters of the lenses are derived from the analytical solutions of the Laplace's equation. The first lens is conceived to produce an overall directive in-phase emission from an array of sources conformed on a cylindrical structure. The second lens allows steering a directive beam to an off-normal direction. Theoretical formulations are given to describe the transformations and full-wave simulations are performed to verify the functionality of the calculated lenses. Theoretical designs are adapted judiciously to be compatible with the constraints of realization technologies so as to have the best agreement between numerical calculations and experimental tests. Prototypes presenting a graded refractive index are fabricated through three-dimensional (3D) polyjet printing using all-dielectric materials and experimental measurements are carried out to validate the proposed lenses. Such easily realizable designs open the way to low-cost all-dielectric broadband microwave lenses for beam forming and collimation.

3.2. In-phase emission restoring lens

The need for restoring in-phase emission is a hot topic since more and more applications require low-profile antennas that are embedded in the skin of vehicles, generally non-planer, so as to reduce aerodynamic drag and fuel consumption. In addition, to be less virtually intrusive, antennas are mounted on non-planar masts. Antennas conformed on cylindrical or spherical surfaces are, therefore, of great interests for communication systems, aerospace applications, high-speed vehicles, rockets and missiles. The influence of the curvature needs to be considered in the design of conformal antennas. While a large curvature radius (compared to the operating wavelength) has little influence on the radiation characteristics, small radius presents the disadvantage of altering the radiation patterns of conformal antennas to a larger extent. For example, antenna elements arranged in an array conformed on a non-planar surface do not radiate in-phase with each other, leading to degradation of the radiated beam and to the performances of the antenna array overall. Due to the varying path lengths of the electromagnetic (EM) waves resulting from the location of the individual antennas on the curved surface, the antenna's radiation patterns are distorted. Therefore, the resulting emission potentially presents multiple beams and a defocusing phenomenon arises. In classical antenna systems, radio-frequency and microwave devices such as phase shifters are used to apply different phase on every individual antenna element to compensate for the introduced phase shifts and, thus, to restore in-phase emissions.

Here, we use the powerful concept of space transformation to calculate the properties of a transformed space supposed to be above a conformal array of antennas that is capable of compensating the conformation characteristics. The transformed space is then translated into the electromagnetic properties of a material that are calculated directly from the geometry of the latter transformed space constituting the lens. In such case, the lens is then able to restore in-phase emission from the antenna array, such that the radiated beam is similar to that of a planar array. The demonstration made here applies to a canonical shape such as a cylinder. However, the conceptual method is more general and can be applied to any shape supporting the radiating elements.

3.2.1. *Theoretical formulations and numerical simulations*

Design of the in-phase emission restoring conformal lens is presented in Figure 3.1 [YI 15a, YI 16b]. Figure 3.1(a) represents the virtual space which is a sector of a cylindrical surface. The physical space presented in Figure 3.1(b) is formed by a 54° angular sector of the annular region between the circles of radius R and r. Segments AB and $\widehat{A'B'}$ are perfect electric conductors, corresponding to the ground plane of the patch antenna array. The segments DA and CB have the same length and are equal to R-r. The transformation established between the physical and virtual spaces aims to transform the conformed patch array into a linear planar one.

As detailed in Chapter 1, the design of a QCTO-based device is associated to the calculation of the 2D Laplace's equation so as to evaluate the coordinate grids deformation during the transformation from an initial virtual space filled with vacuum in the coordinate system (x, y, z) to the real final physical space filled with transformed medium in the coordinate system (x', y', z') [HU 09]. The method implies that the calculation of the electromagnetic material parameters of the transformation medium is equivalent to the computation of spatial deformation field governed by the 2D Laplace's equation with proper boundary conditions, whatever the be shape of the device.

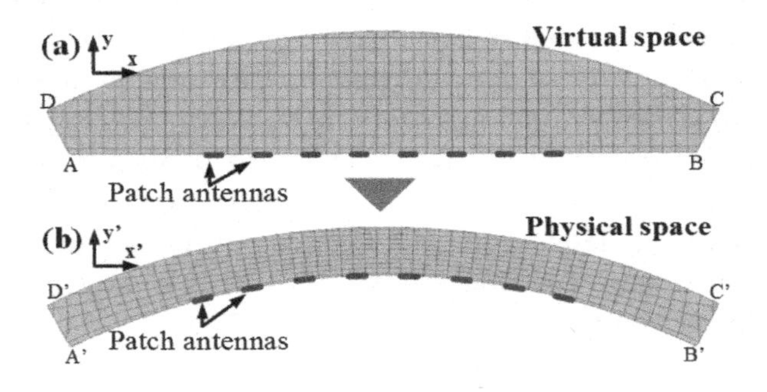

Figure 3.1. *Illustration showing the space mapping from the virtual space to the physical space in the case of the in-phase emission restoring lens [YI 16b]*

In the x-y plane, we suppose that the coordinate transformation between the physical space (x', y') and the virtual space (x, y) is $x = f(x', y')$ and $y = g(x', y')$. The mathematical equivalence of this mapping is expressed by the Jacobian matrix A whose elements are defined by $A = \partial(x,y)/\partial(x',y')$. A is obtained by solving Laplace's equations in the virtual space with respect to specific boundary conditions:

$$\frac{\partial^2 x}{\partial x'^2} + \frac{\partial^2 x}{\partial y'^2} = 0, \quad \frac{\partial^2 y}{\partial x'^2} + \frac{\partial^2 y}{\partial y'^2} = 0 \qquad [3.1]$$

The physical space performs an inverse function of the virtual space. Thus, the Jacobian matrix of this inverse transformation from (x, y) to (x', y') can be represented by A^{-1}. We assume here that the conformal module of the virtual space is 1 while the conformal module of the physical space is M. Once A^{-1} is known, the properties of the intermediate medium can be calculated. In terms of fields' equivalence with the virtual space upon the outer boundaries, Neumann and Dirichlet boundary conditions are set at the edges of the lens. The boundary conditions for the conformal lens are:

$$\begin{aligned} & x\Big|_{B'C',\widehat{C'D'},D'A'} = x', \quad \hat{n}\cdot\nabla x\Big|_{\widehat{A'B'}} = 0 \\ & y\Big|_{\widehat{A'B'}} = 0, \quad y\Big|_{\widehat{C'D'}} = y', \quad \hat{n}\cdot\nabla y\Big|_{B'C',D'A'} = 0 \end{aligned} \qquad [3.2]$$

where $\hat{n}$ is the normal vector to the surface boundaries. The effective property tensors obtained from Laplace's equation are not isotropic in the x-y plane. But, as discussed in Chapter 1, if the conformal module M of the physical space is not quite different with the conformal module of the virtual space, which is 1 in this case, the small anisotropy can be ignored in this case. The near-isotropy resulting from the quasi-conformal mapping leads to such approximation on the anisotropy. For simplicity, the transformation deals with a two-dimensional (2D) model with incident TE-polarized wave having only a z-directed component. Considering the polarization of the excitation, the properties of the intermediate medium can be further simplified as:

$$\varepsilon = \frac{\varepsilon_r}{\det(A^{-1})}, \quad \mu = 1 \qquad [3.3]$$

where ε_r is the permittivity of free space.

A finite element method based on 2D numerical simulations with COMSOL Multiphysics commercial code is used initially to validate the calculated QCTO-based conformal lens numerically. Scattering boundary conditions are set around the computational domain and an array of eight dipole antennas are used as source. The different dipoles are excited with equal amplitude and phase. The electric field of the exciting sources is polarized along the z-direction. The lens is considered to be 2 cm ($A'D' = B'C' = 2$ cm) thick and is parameterized with $R = 29$ cm and $r = 27$ cm.

As shown in Figure 3.2, the permittivity (ε_{zz}) distribution ranges from -0.6 to 2.76 for the calculated conformal lens. Low and high permittivity values are located at the outer borders and at the center of the lens, respectively. The other three components of the permittivity are either close to 1 or close to 0. According to the QCTO concept, this anisotropy can be ignored.

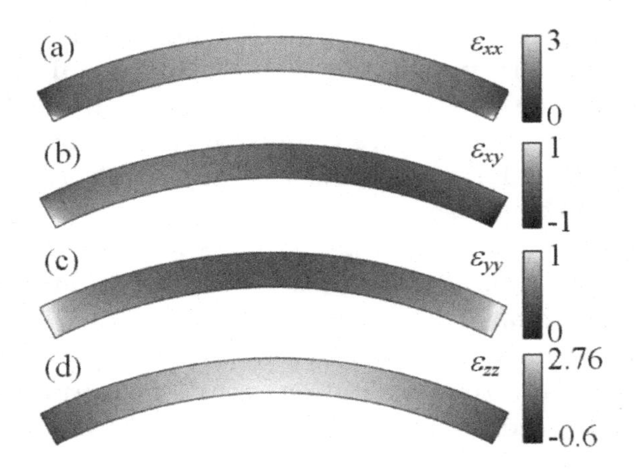

Figure 3.2. *Permittivity distribution for the calculated in-phase emission restoring lens: a) ε_{xx}; b) ε_{xy}; c) ε_{yy}; d) ε_{zz}*

The electric field distribution for different configurations is shown at 10 GHz in Figure 3.3. Three configurations are considered: an ideal rectilinear (planar) dipole array, a conformal dipole array and a dipole array-lens system. As it can clearly be observed in Figure 3.3(a), the outgoing waves of the planar array present planar wavefronts and, therefore, a directive emission. However, in the case of the conformal array alone, the wavefronts are not planar anymore but rather cylindrical, as illustrated in Figure 3.3(b). A defocusing phenomenon is produced, with an electromagnetic radiation producing several beams. This phenomenon is due to the fact that the different dipoles on the cylindrical structure though excited with equal amplitude and phase, do not emit in phase with each other to create a constructive interference that produces a clear focused radiated beam at boresight. The conformal array alone presents a wider beam with a lower intensity as compared to the planar array. This defocusing is corrected when the anisotropic transformation based lens is placed above the conformal array (Figure 3.3(c)). The wavefronts of the conformal array, in presence of the lens, are quasi-planar and agree well with those of the planar array. The performances demonstrate clearly the usefulness of the lens in restoring in-phase emissions. In Figure 3.3(d), the lens is assigned with isotropic material parameters. As with the anisotropic lens, a clear main beam without defocusing is obtained. However, the level of the sidelobes tends to increase, which is due to the fact that off-diagonal components in the material parameter tensor have been neglected for the isotropic lens, as illustrated by the 2D far-field patterns in Figure 3.4.

3.2.2. *3D design, implementation and full-wave simulations*

The calculated isotropic material parameters present a continuous variation of permittivity along the *z*-axis (E-field direction). This is not easy to achieve in practical implementations and, in general, it is necessary to carry out a discrete variation. A method that consists of discretizing the permittivity profile into several zones of different permittivity values is adopted. This discretization secures the characteristics of the ε_{zz} response of the lens.

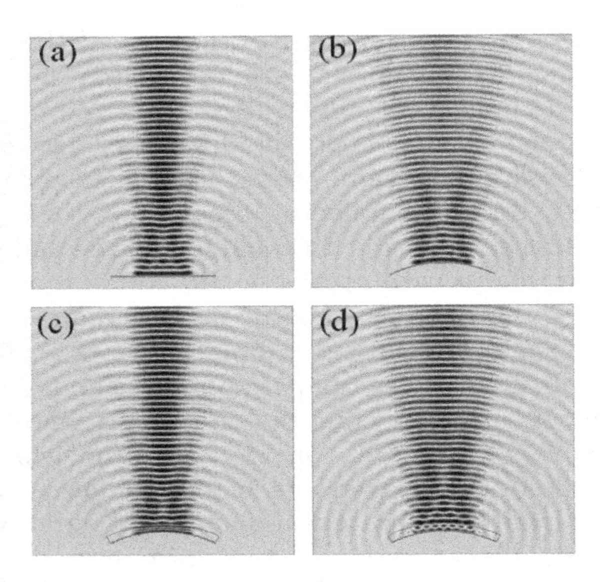

Figure 3.3. *Electric field distribution at 10 GHz: a) planar antenna array alone with a clear main radiated beam; b) conformal antenna array alone showing a defocusing phenomenon; c) conformal antenna array with anisotropic transformed lens; d) conformal antenna array with isotropic transformed lens. The defocusing phenomenon introduced by the conformal array is corrected through the use of the lens*

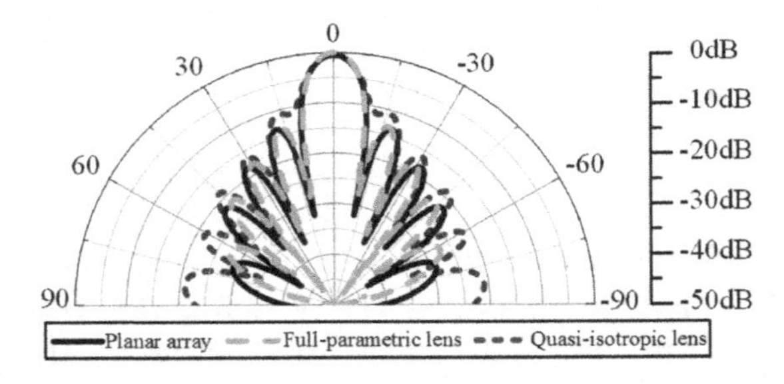

Figure 3.4. *2D far field radiation patterns at 10 GHz illustrating the effect of ignoring anisotropy in the material parameters*

According to effective medium theory, if the operating wavelength is large enough with respect to the size, the composite material can be considered isotropic and homogeneous. A discrete lens model composed of 92 unit cells is proposed. The lens is designed such that the respective permittivity of each cell is considered to be constant across the cell and is equal to the average permittivity within the cell. As illustrated in Figure 3.5, the permittivity ε_{zz} values range from 1.5 to 2.6 in the discrete approximation of the conformal lens. The two edges with permittivity values below 1.5 are suppressed from the continuous lens such that the permittivity ranges between 1.5 and 2.6.

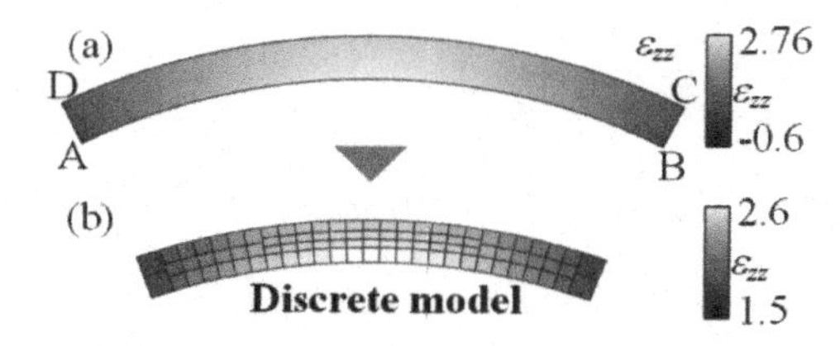

Figure 3.5. *Parameter reduction for the discrete model of the in-phase emission restoring lens*

As only ε_{zz} varies from 1.5 to 2.6 in the lens design, we are able to consider using it over a broad frequency range. The lens is thus realized from non-resonant cells. Air holes in a dielectric host medium of relative permittivity $\varepsilon_h = 2.8$ is, therefore, considered as for the practical implementation, we will consider a dielectric photopolymer with $\varepsilon_r = 2.8$ in a polyjet 3D printing fabrication facility. Suppose that two materials are mixed together, the effective parameter can be calculated by:

$$\varepsilon_{eff} = \varepsilon_a f_a + \varepsilon_h f_h \quad [3.4]$$

where $\varepsilon_a = 1$ and f_a and $f_h = 1 - f_a$ are the volume fraction of the air holes and the host material, respectively. By judiciously adjusting the volume fraction of the air holes through the geometrical dimensions p and r_a, the effective permittivity of the cell can then be engineered as shown in

Figure 3.6. Values ranging from close to 1 to 2.8 can, thus, be obtained for ε_{zz} over a wide frequency-band.

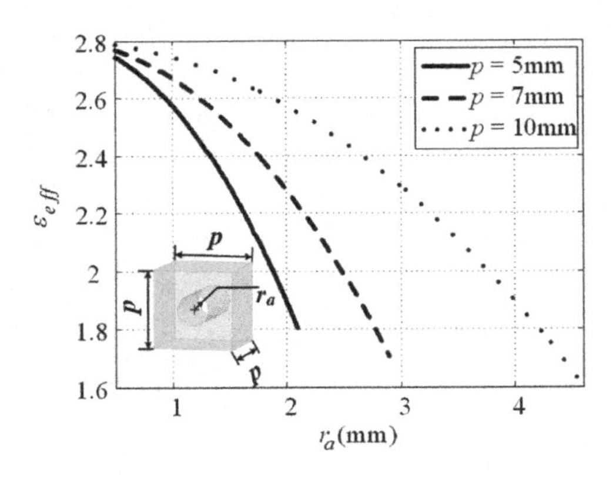

Figure 3.6. *Effective permittivity of a cubic cell composed of air hole in a dielectric host medium. A parametric analysis is performed to extract the effective permittivity value according to the radius r_a of the air hole with respect to a cubic cell of period p (p changes according to the regions of the discrete lens)*

Full-wave simulations using commercial electromagnetic solver HFSS from Ansys are performed to verify the functionality of the lens numerically. A microstrip patch antenna array composed of eight linear radiating elements conformed on a cylindrical surface is considered as the excitation source for the in-phase emission restoring lens as shown in Figure 3.7(a) and the schematic design of the lens is presented in Figure 3.7(b). The lens comprises of five regions with a total of 92 cells where ε_{zz} varies from 1.5 to 2.6 and the dimensions are set as: r = 27 cm and d = 2 cm. Region 2 and region 3 share similar characteristics with respectively region 4 and region 5. Region 1 is discretized into 13 rows of 4 cells with p = 5 mm, regions 2 and 4 are each discretized into 6 rows of 3 cells with p = 6.7 mm and regions 3 and 5 are decomposed each into 1 row of 2 cells with p = 10 mm.

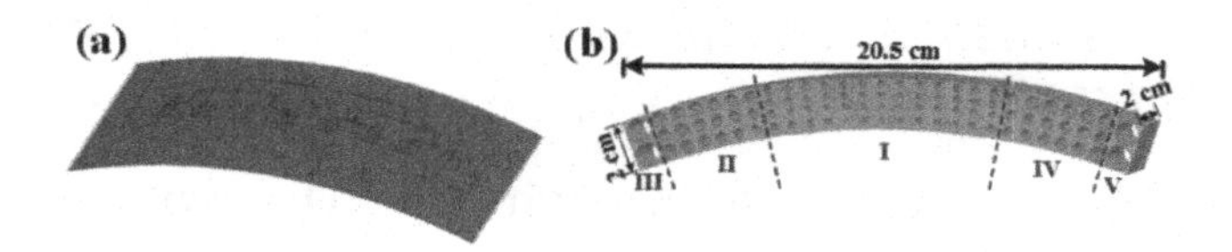

Figure 3.7. *a) Schematic design of the conformal patch array source; b) perspective view of the discrete lens*

The simulated 2D far-field radiation diagrams are presented in Figure 3.8. The planar array presents a directive radiation with a clear narrow main lobe, while the conformal array presents a wider main beam with a lower magnitude. The lens antenna system is able to increase the magnitude of the first lobe and decreases the level of the second lobe. The conformal array without lens has a directivity of 12.3 dB, while the conformal array in presence of lens has a directivity of 15.7 dB, which is better than the directivity of the planar array alone (15.3 dB). Such results confirm the fact that the lens is able to create a constructive interference between the emissions of the radiating elements of the conformal array and, hence, produce an overall in-phase emission. However, though the effect of restoring in-phase emission is clear on the main lobe of the radiation pattern, sidelobes level tends to be high in presence of the lens. This is due to the fact that the anisotropy in permittivity has been ignored, as discussed previously.

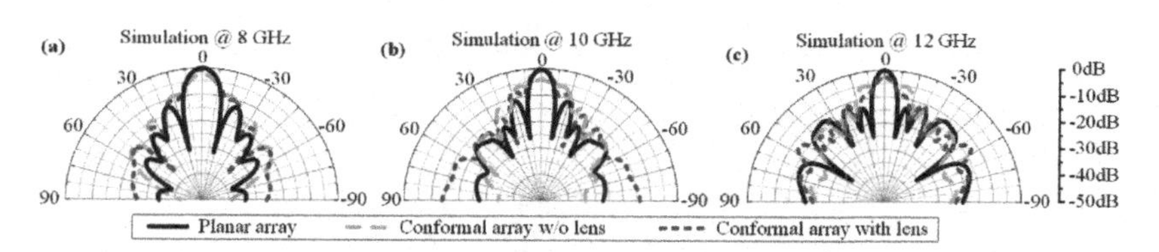

Figure 3.8. *Simulated radiation patterns in the focusing plane (x-y plane); a) 8 GHz; b) 10 GHz; c) 12 GHz. The conformal array presents a distorted diagram with a lower radiation level than the planar array. The lens above the conformal array allows restoring the in-phase emission to create a radiation pattern with a clear directive main beam, similar to a planar array*

3.2.3. *Experimental validation of fabricated in-phase emission restoring lens*

The lens is fabricated through the 3D polyjet printing. The measured electric field mappings of the conformal focusing lens antenna and the measured antenna radiation patterns are depicted at different tested frequencies in 8 GHz to 12 GHz band in Figure 3.9. Planar wavefronts are observed for the lens antenna system, which are consistent with calculated data, indicating a directive radiated beam without defocusing. When associated to the lens, the cylindrical wavefronts of the conformal array of radiators are flattened and transformed into quasi-planar wavefronts and the field distributions are comparable to those of a planar array. The measured far-field antenna patterns clearly show an enhancement in directivity magnitude and a clear main lobe with the reduction of distortion presented by the conformal array alone.

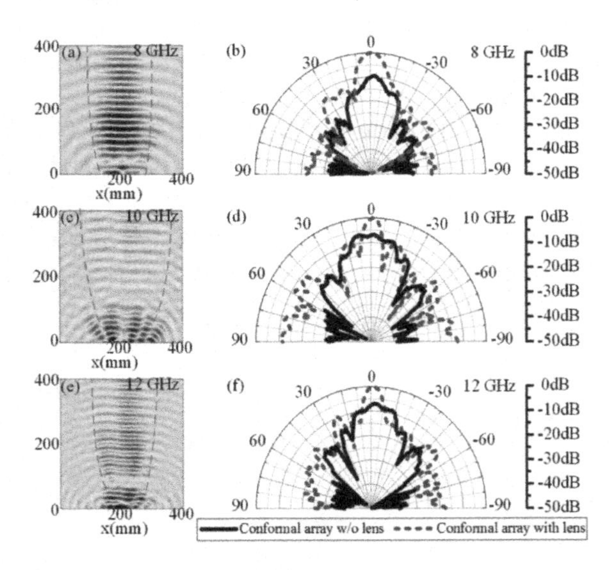

Figure 3.9. *Measured electric near-field distributions and far-field radiation patterns of the conformal lens antenna in the 8–12 GHz frequency band. It can be clearly observed that the designed lens enhances the directivity of the main beam*

3.3. Beam steering lens

Beam steering is a functionality widely used in communication systems to modify propagation direction or to track signals. Such a capability is generally achieved electronically in a phased array antenna which is composed of several radiating elements, each with a phase shifter. Beams are formed by shifting the phase of the signal emitted from each radiating element, to provide constructive/ destructive interference so as to steer the beams in the desired direction [BAL 97]. With the advent of metamaterials and transformation optics concept, we are able to propose novel low-cost lenses as possible alternatives to complex phase shifting systems.

3.3.1. *Theoretical formulations and numerical simulations*

The virtual space which is free space and the physical space which is the transformed medium lens are shown in the schematic principle presented in Figure 3.10. To modify the direction of propagation from normal (along y-axis) to an off-normal direction, we need to transform the angle between the wave vector k and the y-axis. For simplicity, we aim to have a 45° steering. In the virtual space, the segment AB must be inclined at 45° with respect to the y-axis and a radiating element placed on AB will, therefore, produce an emission at 45° off the normal direction. To achieve the same beam steering from a radiating element placed horizontally, we use a transformed lens in the real physical space [YI 15b, YI 16a].

The points B', C' and D' in the physical space share the same locations as B, C and D in the virtual space, respectively. We consider the length of the segment CD to be equal to W and that of BC to be H. Therefore, the segment $A'B'$ has a length $W/\cos(\pi/4)$ and $D'A'$ has a length $W+H$. The determination of the mapping is introduced by solving Laplace's equations subject to the predefined boundary conditions:

$$x\big|_{B'C',C'D',D'A'} = x', \quad \hat{n}\cdot\nabla x\big|_{A'B'} = 0$$
$$y\big|_{B'C',C'D'} = y', \quad y\big|_{A'B'} = \tan\left(\frac{\pi}{4}\right) * \left(x' - \frac{W}{2}\right), \quad \hat{n}\cdot\nabla y\big|_{D'A'} = 0 \qquad [3.5]$$

where $\hat{n}$ is the normal vector to the surface boundaries.

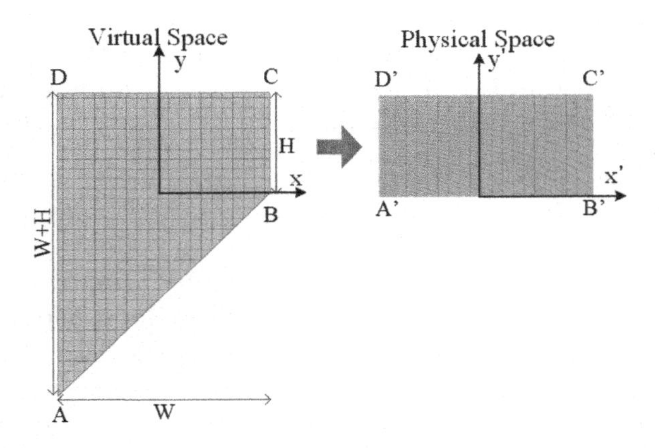

Figure 3.10. *Illustration of the space mapping from the virtual space (radiating element inclined at 45° from the normal direction) to the physical space (horizontally placed radiating element and lens)*

As for the in-phase emission restoring lens, the effective property tensors obtained from Laplace's equation are not isotropic in the x-y plane. For simplicity, the transformation deals with a two-dimensional (2D) model with incident TE-polarized wave having only a z-directed component.

As shown in Figure 3.11, the permittivity (ε_{zz}) distribution ranges from 1 to 6.4 for the calculated beam steering lens. According to the QCTO concept, anisotropy in the material tensor is ignored.

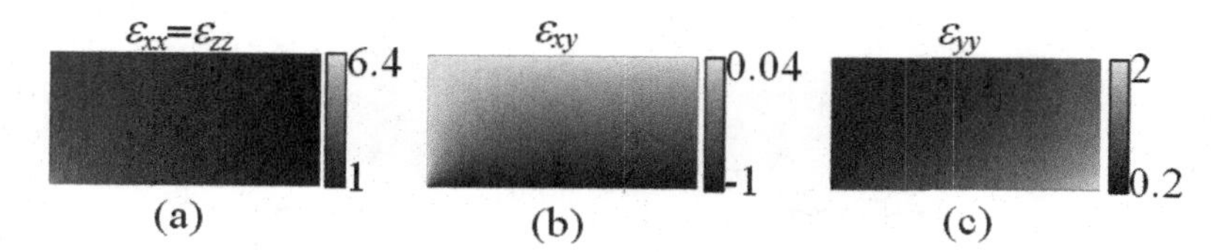

Figure 3.11. *Permittivity distribution for the calculated beam steering lens: a)* $\varepsilon_{xx} = \varepsilon_{zz}$*; b)* ε_{xy}*; c)* ε_{yy}

The electric field distribution for different configurations is shown at 10 GHz in Figure 3.12. In Figure 3.12(a), a line source of length $Ls/\sqrt{2}$ is placed on the inclined boundary *AB*. The whole virtual space is filled with vacuum. As it can be observed clearly, the source emits a beam towards 45° from the vertical direction. On the contrary, a source of length *L* is placed on the boundary *A'B'* in vacuum radiates a beam at boresight (normal) direction, as illustrated in Figure 3.12(b). The anisotropic transformed lens (Figure 3.12(c)) produces the same field distribution as in the initial virtual space (Figure 3.12(a)) after the beam emits out of the boundary *C'D'*. By applying the material properties obtained from transformation calculations, the radiated beam is deflected by 45° after propagating through the lens, as it can be clearly observed in Figure 3.12(c). The lens is, then, assigned isotropic material properties after the parameter reduction process is performed by ignoring the anisotropy in permittivity and keeping only the component ε_{zz} while the permeability is isotropic and equal to 1. It is clear that, since the physical space of the lens is distorted and the conformal module is quite different from 1, the lens still steers the beam but by an angle of 28° (Figure 3.12(d)), which is smaller than the expected 45°. This is due to the fact that the conformal module is much larger than 1. Therefore, anisotropy cannot be ignored if a deflection of 45° is still desired. Such effect is clearly shown by the 2D far field radiation patterns in Figure 3.13.

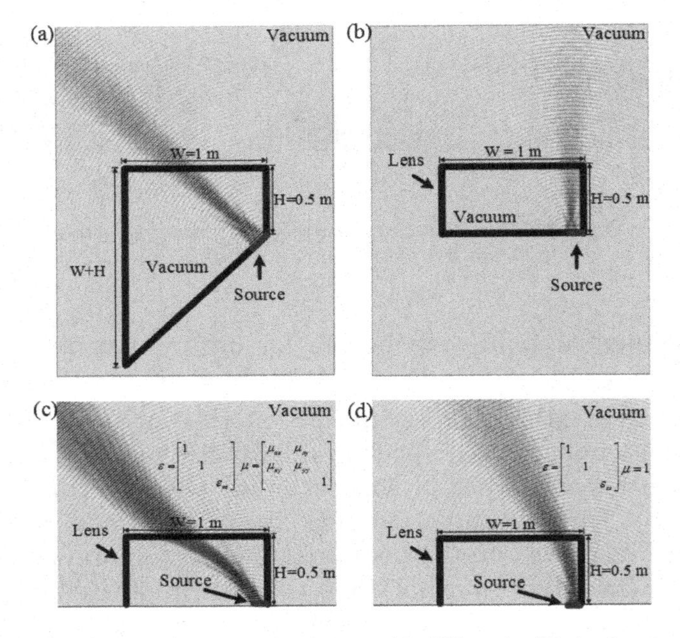

Figure 3.12. *Electric field distribution at 10 GHz: a) 45° inclined antenna source of length* $Ls/\sqrt{2}$ *alone; b) planar antenna source of length Ls alone. c) planar antenna source (length Ls) with anisotropic transformed lens. d) planar antenna source (length Ls) with isotropic lens. The radiated beam is steered to 45° off-normal in presence of the anisotropic lens and 28° in presence of the isotropic lens*

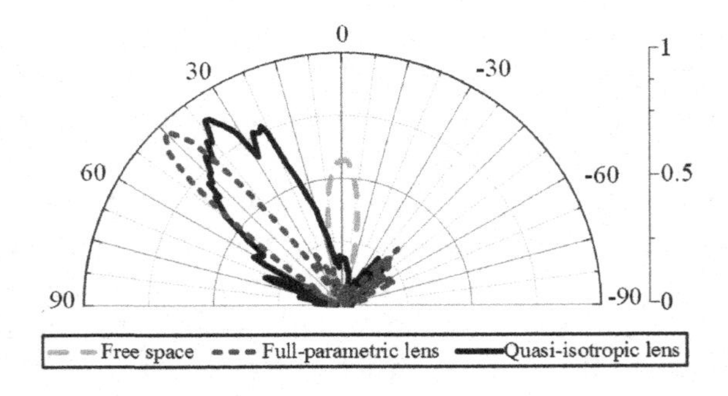

Figure 3.13. *2D far field radiation patterns at 10 GHz illustrating the effect of ignoring anisotropy in the material parameters of the beam steering lens*

3.3.2. *3D design, implementation and full-wave simulations*

The lens is discretized into 170 cells. As illustrated in Figure 3.14, the permittivity ε_{zz} values range from 1 to 2.8 in the discrete approximation of the lens. The edges with permittivity values higher than 2.8 are suppressed in the discrete lens version to accommodate with the dielectric photopolymer with $\varepsilon_r = 2.8$.

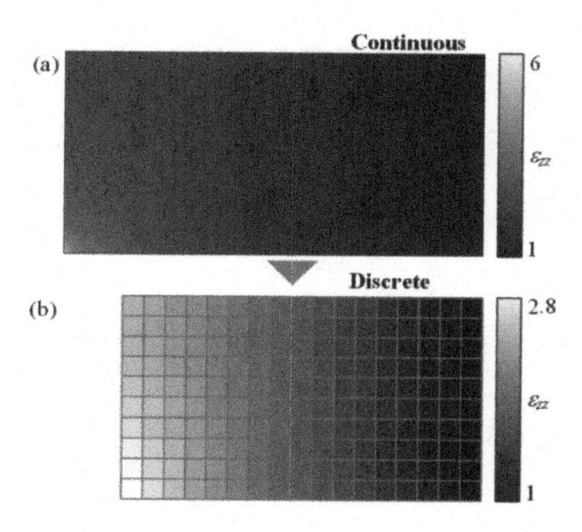

Figure 3.14. *Parameter reduction for the discrete model of the beam steering lens*

Full-wave simulations are performed to numerically verify the functionality of the lens. A microstrip patch antenna array, composed of four equally fed linear radiating elements, is used as primary source (Figure 3.15(a)) and the schematic design of the lens is presented in Figure 3.15(b).

The simulated 3D far-field radiation diagrams are represented in Figure 3.16. The planar array presents a sectorial beam, i.e. a wide beam in one plane and a narrow beam in the other one is obtained since we are using a linear array of radiating elements. As can be observed in Figures 3.16(b), (c) and (d), in presence of the designed 3D dielectric lens, the radiated wavefronts undergo a deflection of 28°, confirming the 2D simulation results and the fact that the lens is able to modify the direction of wave propagation. Moreover, the lens is able to enhance the directivity of the radiated beam from 16 to more than 31.

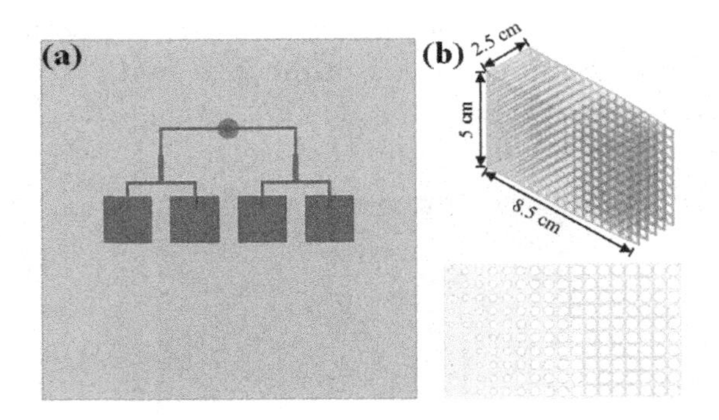

Figure 3.15. *a) Schematic design of the patch array source; b) perspective view of the discrete lens*

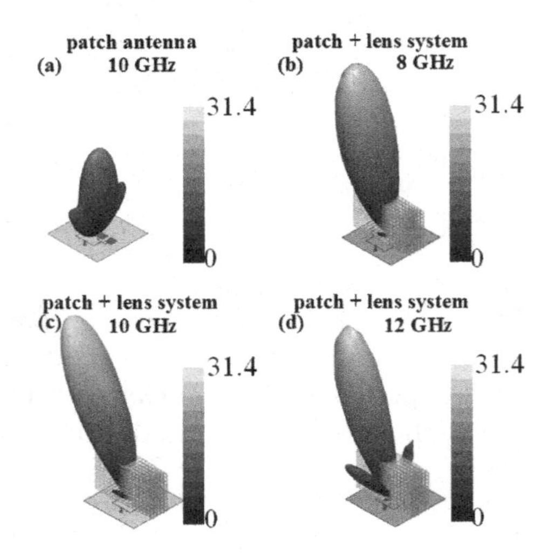

Figure 3.16. *Simulated 3D radiation patterns in linear scale: a) linear array of patch elements at 10 GHz; b) lens antenna system at 8 GHz; c) lens antenna system at 10 GHz; d) lens antenna system at 12 GHz. The influence of the lens is twofold: firstly to enhance the directivity of the patch array source and secondly, to steer the radiated beam*

3.3.3. *Experimental validation of fabricated beam steering lens*

The measured electric field mappings of the beam steering lens antenna and the measured antenna radiation patterns are depicted at different tested frequencies in the 8 GHz to 12 GHz band in Figure 3.17. As predicted by numerical simulations, beam steering behavior is confirmed in measurements. From the different plots, a clear directive main lobe is observed at all tested frequencies for the lens antenna system. The wavefronts emanating from the lens are deflected from the normal direction. Broadband characteristics of the lens are observed due to the use of non-resonant dielectric materials.

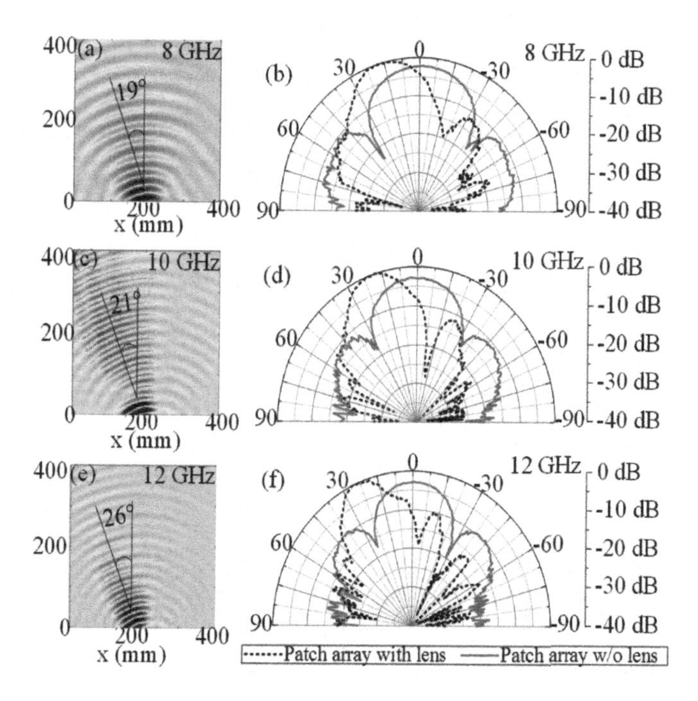

Figure 3.17. *Measured electric near-field distributions and far-field radiation patterns of the beam steering lens antenna in the 8-12 GHz frequency band*

3.4. Conclusion

This chapter is devoted to the design of lenses via space transformation. Two lenses are designed, each modifying wave propagation in a specific way. The lenses are capable of either restoring in-phase emissions from a conformal array of radiators or steering the beam radiated by an antenna. The lenses are implemented using all-dielectric materials allowing them to show broadband performances. The developed lenses can be easily fabricated and reproduced at comparatively low costs using widely accessible technology including 3D printing and present potential airborne and trainborne applications in communication systems.

Conclusion

This book is devoted to the study of the transformation optics concept for the design of novel types of new electromagnetic devices. Transformation optics is an efficient tool to manipulate electromagnetic field distributions. The form-invariance of Maxwell's equations is exploited to design complex electromagnetic media with desired and particular properties. Although this technology makes it possible to generate valid materials parameters, theoretically regardless of frequency, we restrict ourselves to the microwave domain. The technique of transformation optics has no limit in the design, and applications are transposable to any frequency. The applicational perspectives are, therefore, various and in this book several applications for antennas are presented.

Coordinate transformation is able to compress, rotate or fold the space at will. It is applied first to validate the design of an isotropic antenna experimentally by stretching the space around a directive source. Secondly, we show how to design a transformation-based shell that can be used efficiently to recover the radiation diagrams of electromagnetic sources after their miniaturization, therefore enabling isotropic patterns to transform into directive ones. Finally, we present the transformation of a single beam antenna into a multi-beam antenna. The material properties designed from coordinate transformation are always complex or singular. In most cases, the parameter tensor consists of non-diagonal components with values that are either negative or too high to be achieved. Generally, resonant metamaterials are required, which limit the frequency bandwidth of the devices.

Spatial transformation is another method to reshape the electromagnetic field. Other than mapping the virtual and physical space from point to point as in coordinate transformation, the mapping is established by solving Laplace's equation with specific boundary conditions. A conformal beam focusing lens and a beam steering lens are designed by this concept and validated experimentally. Spatial transformation provides a much easier way to design devices. Indeed, in the case of quasi-conformal transformation optics, an in-plane isotropic material realized by non-resonant all-dielectric cells can be applied to the designs, allowing broadband operations. However, this concept requires the virtual and physical spaces to be quite similar to one another, where no subversive mapping can be established, in which case anisotropy cannot be neglected in the material parameters generated.

Transformation optics, therefore, requires the use of complex engineered values of permittivity and permeability. Metamaterials, which are artificially engineered materials, are used to fabricate structures which mimic a material response that has no natural equivalent. Moreover, 3D printing offers a new way to realize complex shapes, if necessary. Existing printers are still limited in the choice of their materials but further developments of new dielectric and/or magnetic composites will improve this technology rapidly and increase the possibility of realizations. Such capabilities lead to an explosion of interests in the physical properties of these materials and the ways in which they can be used to tailor the path of light.

This book is intended to generate a real interest in methods of design and manufacturing processes through the transformation optics approach.

Bibliography

[ALL 09] ALLEN J., KUNDTZ N., ROBERTS D.A. *et al.*, "Electromagnetic source transformations using superellipse equations", *Applied Physics Letters*, vol. 94, p. 194101, 2009.

[BAL 97] BALANIS C.A., *Antenna Theory: Analysis and Design*, 2nd edition, Wiley, New York, 1997.

[BER 08] BERGAMIN L., "Generalized transformation optics from triple space-time metamaterials", *Physical Review A*, vol. 78, p. 43825, 2008.

[CAI 07] CAI W., CHETTIAR U.K., KILDISHEV A.V. *et al.*, "Optical cloaking with metamaterials", *Nature Photonics*, vol. 1, pp. 224–227, 2007.

[CHA 10] CHANG Z., ZHOU X., HU J. *et al.* "Design method for quasiisotropic transformation materials based on inverse Laplace's equation with sliding boundaries", *Optics Express*, vol. 18, pp. 6089–6096, 2010.

[CHE 07] CHEN H., CHAN C.T., "Transformation media that rotate electromagnetic fields", *Applied Physics Letters*, vol. 90, p. 241105, 2007.

[CHE 09a] CHENG Q., MA H.F., CUI T.J., "Broadband planar Luneburg lens based on complementary metamaterials", *Applied Physics Letters*, vol. 95, p. 181901, 2009.

[CHE 09b] CHEN H., HOU B., CHEN S. *et al.*, "Design and experimental realization of a broadband transformation media field rotator at microwave frequencies", *Physical Review Letters*, vol. 102, p. 183903, 2009.

[CHE 10] CHENG Q., CUI T.J., JIANG W.X. *et al.* "An omnidirectional electromagnetic absorber made of metamaterials", *New Journal of Physics*, vol. 12, p. 063006, 2010.

[COU 89] COURANT R., HILBERT D., *Methods of Mathematical Physics*, Wiley-Interscience, vol. 2, 1989.

[CRU 09] CRUDO R.A., O'BRIEN J.G., "Metric approach to transformation optics", *Physical Review A*, vol. 80, p. 33824, 2009.

[DHO 12a] DHOUIBI A., BUROKUR S.N., DE LUSTRAC A. *et al.* "Compact metamaterial-based substrate-integrated Luneburg lens antenna", *IEEE Antennas and Wireless Propagation Letters*, vol. 11, pp. 1504–1507, 2012.

[DHO 12b] DHOUIBI A., BUROKUR S.N., DE LUSTRAC A. *et al.* "Comparison of compact electric-LC resonators for negative permittivity metamaterials", *Microwave and Optical Technology Letters*, vol. 54, pp. 2287–2295, 2012.

[DHO 13a] DHOUIBI A., BUROKUR S.N., DE LUSTRAC A. *et al.* "Low-profile substrate-integrated lens antenna using metamaterials", *IEEE Antennas and Wireless Propagation Letters*, vol. 12, pp. 43–46, 2013.

[DHO 13b] DHOUIBI A., BUROKUR S.N., DE LUSTRAC A. *et al.* "Metamaterial-based half Maxwell fish-eye lens for broadband directive emissions", *Applied Physics Letters*, vol. 102, p. 024102, 2013.

[DOL 05] DOLLING G., ENKRICH C., WEGENER M. *et al.* "Cut-wire pairs and plate pairs as magnetic atoms for optical metamaterials", *Optics Letters*, vol. 30, pp. 3198–3200, 2005.

[DOL 06] DOLLING G., WEGENER M., LINDEN S. *et al.* "Photorealistic images of objects in effective negative-index materials", *Optics Express*, vol. 14, pp. 1842–1849, 2006.

[DRI 06] DRISCOLL T., BASOV D.N., STARR A.F. *et al.*, "Free-space microwave focusing by a negative-index gradient lens", *Applied Physics Letters*, vol. 88, no. 3, p. 081101, 2006.

[ENO 02] ENOCH S., TAYEB G., SABOUROUX P. *et al.* "A metamaterial for directive emission", *Physics Review Letters*, vol. 89, no. 4, p. 213902, 2002.

[ERG 10] ERGIN T., STENGER N., BRENNER P. *et al.* "Three-dimensional invisibility cloak at optical wavelengths", *Science*, vol. 328, p. 337, 2010.

[FAL 04] FALCONE F., LOPETEGI T., LASO M.A.G. *et al.*, "Babinet principle applied to the design of metasurfaces and metamaterials", *Physical Review Letters*, vol. 93, no. 4, p. 197401, 2004.

[FAR 08] FARHAT M., ENOCH S., GUENNEAU S. *et al.* "Broadband cylindrical acoustic cloak for linear surface waves in a fluid", *Physical Review Letters*, vol. 101, p. 134501, 2008.

[GAB 09] GABRIELLI L.H., CARDENAS J., POITRAS C.B. *et al.* "Silicon nanostructure cloak operating at optical frequencies", *Nature Photonics*, vol. 3, pp. 461–463, 2009.

[GAI 08] GAILLOT D.P., CROËNNE C., LIPPENS D. "An all-dielectric route for terahertz cloaking", *Optics Express*, vol. 16, pp. 3986–3992, 2008.

[GEN 09] GENOV D.A., ZHANG S., ZHANG X. "Mimicking celestial mechanics in metamaterials", *Nature Physics*, vol. 5, pp. 687–692, 2009.

[GRE 05] GREEGOR R.B., PARAZZOLI C.G., NIELSEN J.A. *et al.* "Simulation and testing of a graded negative index of refraction lens", *Applied Physics Letters*, vol. 87, no. 3, p. 091114, 2005.

[GRE 07] GREENLEAF A., KURYLEV Y., LASSAS M. *et al.* "Electromagnetic wormholes and virtual magnetic monopoles from metamaterials", *Physical Review Letters*, vol. 99, p. 183901, 2007.

[GRE 08] GREENLEAF A., KURYLEV Y., LASSAS M. *et al.* "Isotropic transformation optics: approximate acoustic and quantum cloaking", *New Journal of Physics*, vol. 10, p. 115024, 2008.

[GUE 05] GUENNEAU S., GRALAK B., PENDRY J.B., "Perfect corner reflector", *Optics Letters*, vol. 30, pp. 1204–1206, 2005.

[HAN 10] HAN T., QIU C.W., "Isotropic nonmagnetic flat cloaks degenerated from homogeneous anisotropic trapeziform cloaks", *Optics Express*, vol. 18, pp. 13038–13043, 2010.

[HU 09] HU J., ZHOU X., HU G., "Design method for electromagnetic cloak with arbitrary shapes based on Laplace's equation", *Optics Express*, vol. 17, pp. 1308–1320, 2009.

[HUA 09] HUANGFU J., XI S., KONG F. *et al.*, "Application of coordinate transformation in bent waveguides", *Journal of Applied Physics*, vol. 104, p. 014502, 2009.

[JIA 08a] JIANG W.X., CUI T.J., CHENG Q. *et al.* "Design of arbitrarily shaped concentrators based on conformally optical transformation of non-uniform rational B-spline surfaces", *Applied Physics Letters*, vol. 92, p. 264101, 2008.

[JIA 08b] JIANG W.X., CUI T.J., MA H.F. *et al.* "Cylindrical-to-plane-wave conversion *via* embedded optical transformation", *Applied Physics Letters*, vol. 92, p. 261903, 2008.

[JIA 11a] JIANG Z.H., GREGORY M.D., WERNER D.H., "A broadband monopole antenna enabled by an ultrathin anisotropic metamaterial coating", *IEEE Antennas and Wireless Propagation Letters*, vol. 10, pp. 1543–1546, 2011.

[JIA 11b] JIANG W.X., CUI T.J., "Radar illusion *via* metamaterials," *Physical Review E*, vol. 83, p. 026601, February 2011.

[JIA 13] JIANG W.X., QIU C.W., HAN T. *et al.* "Creation of ghost illusions using wave dynamics in metamaterials", *Advanced Functional Materials*, vol. 23, pp. 4028–4034, 2013.

[KAD 14] KADIC M., DUPONT G., ENOCH S. *et al.* "Invisible waveguides on metal plates for plasmonic analogs of electromagnetic wormholes", *Physical Review A*, vol. 90, p. 043812, 2014.

[KAF 05] KAFESAKI M., KOSCHNY T., PENCIU R.S. *et al.*, "Left-handed metamaterials: detailed numerical studies of the transmission properties", *Journal of Optics A: Pure and Applied Optics*, vol. 7, pp. S12–S22, 2005.

[KAN 09] KANTÉ B., GERMAIN D., DE LUSTRAC A., "Experimental demonstration of a nonmagnetic metamaterial cloak at microwave frequencies", *Physical Review B*, vol. 80, p. 201104, 2009.

[KON 07] KONG F., WU B.-I., KONG, J.A. *et al.* "Planar focusing antenna design by using coordinate transformation technology", *Applied Physics Letters*, vol. 91, p. 253509, 2007.

[KUN 08] KUNDTZ N., ROBERTS D.A., ALLEN J. *et al.* "Optical source transformations", *Optics Express*, vol. 16, pp. 21215–21222, 2008.

[KUN 10] KUNDTZ N., SMITH D.R., "Extreme-angle broadband metamaterial lens", *Nature Materials*, vol. 9, pp. 129–132, 2010.

[KWO 08] KWON D.H., WERNER D.H., "Transformation optical designs for wave collimators, flat lenses and right-angle bends", *New Journal of Physics*, vol. 10, p. 115023, 2008.

[LAI 09] LAI Y., NG J., CHEN H.Y. *et al.*, "Illusion optics: the optical transformation of an object into another object", *Physical Review Letters*, vol. 102, p. 253902, 2009.

[LAN 09] LANDY N.I., PADILLA W.J., "Guiding light with conformal transformations", *Optics Express*, vol. 17, pp. 14872–14879, 2009.

[LAN 14] LANDY N., URZHUMOV Y., SMITH D.R., "Quasi-conformal approaches for two and three-dimensional transformation optical media", in WERNER D.H., KWON D.-H. (eds.), *Transformation Electromagnetics and Metamaterials*, Springer, pp. 1–32, 2014.

[LEO 00] LEONHARDT U., "Space-time geometry of quantum dielectrics", *Physical Review A*, vol. 62, p. 012111, 2000.

[LEO 06a] LEONHARDT U., "Optical conformal mapping", *Science*, vol. 312, pp. 1777–1780, 2006.

[LEO 06b] LEONHARDT U., "Notes on conformal invisibility devices", *New Journal of Physics*, vol. 8, p. 118, 2006.

[LEO 08] LEONHARDT U., TYC T., "Superantenna made of transformation media", *New Journal of Physics*, vol. 10, p. 115026, 2008.

[LI 08] LI J., PENDRY J.B., "Hiding under the carpet: a new strategy for cloaking", *Physical Review Letters*, vol. 101, p. 203901, 2008.

[LI 10] LI C., MENG X.K., LIU X. *et al.*, "Experimental realization of a circuit-based broadband illusion-optics analogue", *Physical Review Letters*, vol. 105, p. 233906, 2010.

[LIE 11] LIER E., WERNER D.H., SCARBOROUGH C.P. *et al.*, "An octave-bandwidt negligible-loss radiofrequency metamaterial", *Nature Materials*, vol. 10, pp. 216–222, 2011.

[LIU 08] LIU R., CHENG Q., HAND T. *et al.* "Experimental demonstration of electromagnetic tunneling through an epsilon-near-zero metamaterial at microwave frequencies", *Physical Review Letters*, vol. 100, no. 4, p. 023903, 2008.

[LIU 09a] LIU R., CHENG Q., CHIN J.Y. *et al.* "Broadband gradient index microwave quasi-optical elements based on nonresonant metamaterials", *Optics Express*, pp. 21030–21041, 2009.

[LIU 09b] LIU R., JI C., MOCK J.J., *et al.* "Broadband ground-plane cloak", *Science*, vol. 323, pp. 366–369, 2009.

[LIU 09c] LIU R., YANG X.M., GOLLUB J.G. *et al.*, "Gradient index circuit by waveguided metamaterials", *Applied Physics Letters*, vol. 94, p. 073506, 2009.

[LUO 08a] LUO Y., CHEN H., ZHANG J. *et al.*, "Design and analytical full-wave validation of the invisibility cloaks, concentrators, and field rotators created with a general class of transformations", *Physical Review B*, vol. 77, p. 125127, 2008.

[LUO 08b] LUO Y., ZHANG J., RAN L. *et al.* "Controlling the emission of electromagnetic source", *PIERS Online*, vol. 4, pp. 795–800, 2008.

[LUO 08c] LUO Y., ZHANG J., RAN L. *et al.*, "New concept conformal antennas utilizing metamaterial and transformation optics", *IEEE Antennas and Wireless Propagation Letters*, vol. 7, pp. 509–512, 2008.

[MA 09a] MA H.F., CHEN X., XU H.S. *et al.*, "Experiments on high-performance beam-scanning antennas made of gradient-index metamaterials", *Applied Physics Letters*, vol. 95, p. 094107, 2009.

[MA 09b] MA Y.G., ONG C.K., TYC T. *et al.*, "An omnidirectional retroreflector based on the transmutation of dielectric singularities", *Nature Materials*, vol. 8, pp. 639–642, 2009.

[MA 10a] MA H.F., CHEN X., YANG X.M. *et al.*, "A broadband metamaterial cylindrical lens antenna", *Chinese Science Bulletin*, vol. 55, pp. 2066–2070, 2010.

[MA 10b] MA H.F., CUI T.J., "Three-dimensional broadband and broad-angle transformation-optics lens", *Nature Communications*, vol. 1, p. 124, 2010.

[MA 11] MA Y.G., SAHEBDIVAN S., ONG C.K. *et al.*, "Evidence for subwavelength imaging with positive refraction", *New Journal of Physics*, vol. 13, p. 033016, 2011.

[MEI 11] MEI Z.L., BAI J., CUI T.J., "Experimental verification of a broadband planar focusing antenna based on transformation optics", *New Journal of Physics*, vol. 13, p. 063028, 2011.

[NAR 09] NARIMANOV E.E., KILDISHEV A.V., "Optical black hole: Broadband omnidirectional light absorber", *Applied Physics Letters*, vol. 95, p. 041106, 2009.

[NIC 70] NICOLSON A.M., ROSS G.F., "Measurement of the intrinsic properties of materials by time-domain techniques," *IEEE Transactions on. Instrumentation and Measurement*, vol. 19, pp. 377–382, 1970.

[OBR 02] O'BRIEN S., PENDRY J.B., "Photonic band-gap effects and magnetic activity in dielectric composites", *Journal of Physics: Condensed Matter*, vol. 14, pp. 4035–4044, 2002.

[PEN 98] PENDRY J.B., HOLDEN A.J., ROBBINS D.J. *et al.*, "Low frequency plasmons in thin wire structures", *Journal of Physics: Condensed Matter*, vol. 10, pp. 4785–4809, 1998.

[PEN 99] PENDRY J.B., HOLDEN A.J., ROBBINS D.J. *et al.*, "Magnetism from conductors and enhanced nonlinear phenomena",*IEEE Transanctions on Microwave Theory and Techniques*, vol. 47, pp.2075–2084, 1999.

[PEN 06] PENDRY J. B., SCHURIG D., SMITH D.R., "Controlling electromagnetic fields", *Science*, vol. 312, pp. 1780–1782, 2006.

[PFE 10] PFEIFFER C., GRBIC A., "A printed, broadband Luneburg lens antenna", *IEEE Transactions on Antennas and Propagation*, vol. 58, pp. 3055–3059, 2010.

[PLE 60] PLEBANSKI J., "Electromagnetic waves in gravitational fields", *Physical Review*, vol. 118, pp. 1396–1408, 1960.

[POP 09] POPA B.I., ALLEN J., CUMMER S.A., "Conformal array design with transformation electromagnetics", *Applied Physics Letters*, vol. 94, p. 244102, 2009.

[QUE 13] QUEVEDO-TERUEL O., TANG W., MITCHELL-THOMAS R.C. *et al.*, "Transformation optics for antennas: why limit the bandwidth with metamaterials?", *Scientific Reports*, vol. 3, p. 1903, 2013.

[RAH 08a] RAHM M., SCHURIG D., ROBERTS D.A. *et al.*, "Design of electromagnetic cloaks and concentrators using form-invariant coordinate transformations of Maxwell's equations", *Photonics and Nanostructures-Fundamentals and Applications*, vol. 6, pp. 87–95, 2008.

[RAH 08b] RAHM M., CUMMER S.A., SCHURIG D. *et al.*, "Optical design of reflectionless complex media by finite embedded coordinate transformations", *Physical Review Letters*, vol. 100, p. 63903, 2008.

[RAH 08c] RAHM M., ROBERTS D.A., PENDRY J.B. *et al.*, "Transformation-optical design of adaptive beam bends and beam expanders", *Optics express*, vol. 16, pp. 11555–11567, 2008.

[ROB 09] ROBERTS D.A., RAHM M., PENDRY J.B. *et al.*, "Transformation-optical design of sharp waveguide bends and corners", *Applied Physics Letters*, vol. 93, p. 251111, 2009.

[SCH 06a] SCHURIG D., MOCK J.J., JUSTICE B.J. *et al.*, "Metamaterial electromagnetic cloak at microwave frequencies", *Science*, vol. 314, pp. 977–980, 2006.

[SCH 06b] SCHURIG D., MOCK J.J., SMITH D.R., "Electric-field-coupled resonators for negative permittivity metamaterials", *Applied Physics Letters*, vol. 88, no. 3, p. 041109, 2006.

[SCH 10] SCHMIELE M., VARMA V.S., ROCKSTUHL C. *et al.*, "Designing optical elements from isotropic materials by using transformation optics", *Physical Review A*, vol. 81, p. 33837, 2010.

[SHA 07] SHALAEV V.M., "Optical negative-index metamaterials", *Nature Photonics*, vol. 1, pp. 41–48, 2007.

[SHE 01] SHELBY R.A., SMITH D.R., SCHULTZ S., "Experimental verification of a negative index of refraction", *Science*, vol. 292, pp. 77–79, 2001.

[SMI 02] SMITH D.R., SCHULTZ S., MARKOS P. *et al.*, "Determination of effective permittivity and permeability of metamaterials from reflection and transmission coefficients", *Physical Review B*, vol. 65, p. 195104, 2005.

[SMI 05] SMITH D.R., MOCK J.J., STARR A.F. *et al.*, "Gradient index metamaterials", *Physical Review E*, vol. 71, p. 036609, 2005.

[TAM 24] TAMM I.Y., "Electrodynamics of an anisotropic medium in the special theory of relativity", *Russian Journal of Physical Chemistry*, vol. 56, p. 248, 1924.

[TAN 10] TANG W., ARGYROPOULOS C., KALLOS E. *et al.*, "Discrete Coordinate Transformation for Designing All-dielectric Flat Antennas", *IEEE Transactions on Antennas and Propagation*, vol. 58, pp. 3795–3804, 2010.

[TAN 91] TANNERY P., HENRY C., *Œuvres de Pierre de Fermat*, vol. 2, Gauthier–Villars, 1891.

[THO 99] THOMPSON J.F., SONI B.K., WEATHERILL N.P., *Handbook of grid generation*, CRC Press, Boca Raton, 1999.

[THO 11] THOMPSON R.T., CUMMER S.A., FRAUENDIENER J., "A completely covariant approach to transformation optics", *Journal of Optics*, vol. 13, p. 024008, 2011.

[TIC 11] TICHIT P.H., BUROKUR S.N., DE LUSTRAC A., "Transformation media producing quasi-perfect isotropic emission", *Optics Express*, vol. 19, pp. 20551–20556, 2011.

[TIC 13a] TICHIT P.H., BUROKUR S.N., DE LUSTRAC A., "Reducing physical appearance of electromagnetic sources", *Optics Express*, vol. 21, pp. 5053–5062, 2013.

[TIC 13b] TICHIT P.H., BUROKUR S.N., QIU C.-W. *et al.*, "Experimental verification of isotropic radiation from a coherent dipole source via electric-field-driven LC resonator metamaterials", *Physical Review Letters*, vol. 111, p. 133901, 2013.

[TIC 14] TICHIT P.H., BUROKUR S.N., DE LUSTRAC A., "Spiral-like multi-beam emission *via* transformation electromagnetics", *Journal of Applied Physics*, vol. 115, p. 024901, 2014.

[TUR 10] TURPIN J.P., MASSOUD A.T., JIANG Z.H. *et al.*, "Conformal mappings to achieve simple material parameters for transformation optics devices", *Optics Express*, vol. 18, pp. 244–252, 2010.

[VAL 09] VALENTINE J., LI J., ZENTGRAF T. *et al.*, "An optical cloak made of dielectrics", *Nature Materials*, vol. 8, pp. 568–571, 2009.

[WEI 74] WEIR W.B., "Automatic measurement of complex dielectric constant and permeability at microwave frequencies", *Proceedings of the IEEE*, vol. 62, pp. 33–36, 1974.

[YAN 08] YAN W., YAN M., RUAN Z. *et al.*, "Coordinate transformations make perfect invisibility cloaks with arbitrary shape", *New Journal of Physics*, vol. 10, p. 043040, 2008.

[YI 15a] YI J., BUROKUR S.N., PIAU G.P., DE LUSTRAC A., "Restoring in-phase emissions from non-planar radiating elements using a transformation optics based lens", *Applied Physics Letters*, vol. 107, p. 024101, 2015.

[YI 15b] YI J., BUROKUR S.N., DE LUSTRAC A., "Conceptual design of a beam steering lens through transformation electromagnetics", *Optics Express*, vol. 23, pp. 12942–12951, 2015.

[YI 16a] YI J., BUROKUR S.N., PIAU G.P. *et al.*, "Coherent beam control with an all-dielectric transformation optics based lens", *Scientific Reports*, vol. 6, p. 18819, 2016.

[YI 16b] YI J., BUROKUR S.N., PIAU G.P. *et al.*, "3D printed broadband transformation optics based all-dielectric microwave lenses", *Journal of Optics*, in press, 2016.

[ZHA 09] ZHANG J., LUO Y., CHEN H. *et al.*, "Guiding waves through an invisible tunnel", *Optics Express*, vol. 17, pp. 6203–6208, 2009.

Index

Printed in the United States
By Bookmasters